AGRICULTURE IN FRANCE
ON THE EVE OF THE RAILWAY AGE

CROOM HELM HISTORICAL GEOGRAPHY SERIES
Edited by R.A. Butlin

THE DEVELOPMENT OF THE IRISH TOWN
Edited by R.A. Butlin

THE MAKING OF URBAN SCOTLAND
I.H. Adams

THE FUR TRADE OF THE AMERICAN WEST, 1807-1840
David J. Wishart

THE MAKING OF THE SCOTTISH COUNTRYSIDE
Edited by M.L. Parry and T.R. Slater

LORD AND PEASANT IN NINETEENTH-CENTURY BRITAIN
Dennis R. Mills

Agriculture in France on the Eve of the Railway Age

HUGH CLOUT

CROOM HELM LONDON

BARNES & NOBLE BOOKS
TOTOWA, NEW JERSEY

©1980 Hugh Clout
Croom Helm Ltd, 2-10, St John's Road, London SW11

British Library Cataloguing in Publication Data

Clout, Hugh Donald
    Agriculture in France on the eve of the railway age. — (Croom Helm
    historical geography series).
    1. Agriculture — Economic aspects — France — History — 19th century
    I. Title
    338.1'0944   HD1945

ISBN 0-85664-919-8

First published in the USA 1980 by
BARNES & NOBLE BOOKS
81 ADAMS DRIVE
TOTOWA, NEW JERSEY, 07512
ISBN: 0-389-20017-4

Printed and bound in Great Britain

CONTENTS

TABLES

FIGURES

ABBREVIATIONS

Journal Titles

AAF	Annales de l'Agriculture Française
AASAF	Annales Administratives et Statistiques de l'Agriculture
ACSS	Actes du Congrès des Sociétés Savantes
ADH	Annales de Démographie Historique
AE	Annales de l'Est
AESC	Annales Economies, Sociétés, Civilisations
AG	Annales de Géographie
AH	Agricultural History
AHR	Agricultural History Review
AN	Annales de Normandie
AOF	Agriculture de l'Ouest de la France
APAPER	Annales Provençales d'Agriculture et d'Economie Rurale

AS	Annales de Statistique
ASLIA	Annales Scientifiques, Littéraires et Industrielles de l'Auvergne
CH	Cahiers d'Histoire
ER	Etudes Rurales
FHS	French Historical Studies
ISEA	Institut des Sciences Economiques Appliquées
JAP	Journal d'Agriculture Pratique
JCH	Journal of Contemporary History
JEEH	Journal of European Economic History
JEH	Journal of Economic History
JMH	Journal of Modern History
P	Population
RGE	Revue Géographique de l'Est
RGPSO	Revue Géographique des Pyrénées et du Sud-Ouest
RH	Revue Historique
RHES	Revue d'Histoire Economique et Sociale
TIBG	Transactions, Institute of British Geographers

Archival Abbreviations

AD	Archives Départementales
AN	Archives Nationales
BNCC	Bibliothèque Nationale: Collection Cartographique

A Note on Placenames

Names of *départements* are given in their nineteenth-century form (e.g. 'Charente-Inférieure' rather than 'Charente-Maritime') but where English versions exist for cities (e.g. 'Lyons'), islands (e.g. 'Corsica'), *pays* (e.g. 'Upper Normandy'), provinces (e.g. 'Brittany'), regions (e.g. 'Paris Basin'), rivers (e.g. 'the Rhine') and other geographic phenomena they have been preferred. Unlike the French convention, the preceding definite article has normally been omitted before the names of départements and pays.

PREFACE

This book was born out of professional, academic and personal frustration. As a university teacher and no more than a part-time researcher I had written a number of rather disparate articles which reflect prime commitment to teaching, by the written and the spoken word, and perceived limitations on research time. I needed to undertake a more sustained project. As a historical geographer, at least some of the time, I had been challenged by the extreme difficulty, even near impossibility, of charting rather than inferring social and economic processes in nineteenth-century France and in so many other contexts. In the present book I fight shy of that problem and attempt to reconstruct conditions at one specific time, not seeing that task as an end in itself or because of blind faith in quantitative cartography but in the hope that it will provide a datum plane which may be of interest to process-orientated researchers in the future.

Over the years I have received financial support for work in France from the Centre National de la Recherche Scientifique, the Social Science Research Council, the Central Research Fund (University of London) and University College London. The present book and a number of earlier publications have been made possible in this way. My Head of Department, Professor Bill Mead, has provided every encouragement for me to maintain the 'French connection'. To Hugh Prince I owe a special debt of gratitude for kindling my academic interest in France and for monitoring all subsequent phases in my work, even to the point of critically reading an earlier draft of this book. Few resident members of the Maconochie Foundation have escaped my questioning on many relevant issues about which I was woefully ignorant. I thank them for their interest and enlightenment. In particular I extend my gratitude to members of the 'Monday group' 1977-8 without whose support and understanding this book would not have been completed. Outside the group, Bill Campbell provided moral support and invaluable technical assistance in generating the computergraphics. Only a small sample of his hard work is included in the present book. Rick Davidson and especially Christine Daniels shouldered a formidable cartographic burden during the early days when my desk calculator was producing results more speedily than the computer. Chris Cromarty photographed a large number of line drawings and developed new skills in handling micro-filmed computergraphics. Annabel Swindells, Claudette John and their colleagues typed and retyped so much, coping splendidly with my handwriting, unprofessional typing and mutterings into the dictaphone. My thanks to all of them for giving so much of their time and expertise.

Librarians and archivists in London, Paris and a number of French provincial cities helped me in many ways. Special thanks are due to the staff of the Official Publications Library (British Library) for their friendliness and to fellow readers for their idiosyncracies which can enliven the dullest afternoon and the most tedious volume. My tutorial students and my non-academic friends have offered gentle criticisms of my work and made me think a great deal. Malcolm and Wyn helped me keep body and soul together and my mother showed her skills as a human computer and graphical plotter with an amazing degree of enthusiasm. Words cannot express my gratitude to each of them. As a countryman born and bred and a lifelong gardener my father had much practical advice to offer on rural life in years past. He died while this book was in preparation and so many of his experiences must remain unshared — to my profound regret.

Agriculture and Industry

In spite of a rich inheritance of urban life and traditional manufactures, nineteenth-century France moved more slowly towards factory-based industrialisation than did her neighbours across the Channel and across the Rhine. Their national populations increased rapidly throughout the century but in France rates of growth declined substantially after 1850 with more deaths than births being recorded in some years. Unlike her neighbours, France retained a surprisingly large population in the countryside and a large workforce on the land. These facts were the root cause of the slowness of her economic and social transformation. By the turn of the century 59 per cent of the French nation still lived in rural areas and 42 per cent of the workforce was engaged in farming and forestry, by comparison with 46 per cent and 31 per cent in Germany, and only 25 per cent and 12 per cent in Great Britain. As Napoleon had recognised a century earlier, agriculture was the 'heart' of the French economy.[1] Agrarian interests retained great significance in French political life and the average Frenchman of 1900 was still unquestionably a countryman. Many 'Parisians' had been born in the countryside and certainly kept their roots implanted in their home village and *pays natal*. In spite of highly centralised systems of administration and education and a railway network which linked each pays to its *chef-lieu* and that in turn to Paris, the French countryside continued to display a great diversity of agricultural activities, settlements and lifestyles which mirrored the enormity of physical and human resources that the 'hexagon' contained. It was in no way fortuitous that several generations of French geographers devoted so much of their energy to recording, classifying and trying to explain the complexity of the rural environment.[2]

Historians interested in the evolution of national economies approached the study of nineteenth-century France armed with a different set of questions. Why, for example, did France appear to be such a laggard in the race for industrialisation and urban growth? How did the primary and secondary sectors of the economy evolve and what was the nature of their relationship? Was there an 'industrial revolution' or, indeed, an 'agricultural revolution'? These and allied topics have exercised economic historians on both sides of the Atlantic, albeit in rather different ways. French scholars have adopted at least three types of approach. Some have analysed the performance of particular industries or financial institutions and usually framed their research in the context of individual regions rather than the whole nation. Others have sought to trace the evolution of complete regional economies over defined spans of time rather than concentrating specifically on the industrial sector. Members of the Paris-based school of 'quantitative history' have adopted a third approach by using national accounts to devise indicators of economic performance from the eighteenth century to the very recent past.[3] Their methods of statistical interpolation are ingenious and their conclusions often controversial, but such work provides valuable national statements for chosen sectors of economic life.

A number of American and British scholars have tried to synthesise this enormous detail of fundamental research for presentation to an international audience. Retardation of economic growth in France has been attributed to many factors, including entrepreneurial timidity and inadequate domestic supplies of coal and other raw materials, but attention has always been brought back to the underlying characteristics of land ownership and agricultural production which retained so great a proportion of the workforce for so long a time. In an early paper, Clough argued that agricultural techniques remained poor and levels of farm investment low because the land reforms of the French Revolution and subsequent inheritance laws helped make France a land of predominantly small, individually-owned holdings.[4] Peasant proprietorship was bound up with agricultural underemployment, prevented growth in farm size and inhibited use of advanced techniques.[5] Fiscal protection later in the nineteenth century shielded this large and relatively inefficient farming sector from competition and subsequent change. Labour and capital remained in agriculture rather than being transferred from it as a prerequisite for industrial growth. The net result was economic retardation by comparison with accelerating expansion in Germany and Great Britain. In the past two decades statistical manipulation has become more sophisticated and conclusions have been more qualified. O'Brien and Keyder have insisted that it is not illuminating to assume that the English 'path to the twentieth century' constituted the best or the normal practice and that France was particularly retarded, since each nation was placed in and responded to its own economic, social and political context.[6] Nonetheless, national differences in property ownership and agricultural employment were recognised as fundamental.

In their quest to monitor and explain economic circumstances during the nineteenth century French historians have devoted particular attention to manufacturing. Such emphasis is hardly surprising since industrialisation and urbanisation formed the most striking manifestation of modernisation in nineteenth century Europe. It is, however, ironic that relatively little attention has been paid to rural life since so many historians have attributed at least part of their explanations for slow industrial growth to the state of the agricultural realm. Admittedly some enquiries by historians have examined 'total' regional economies in the nineteenth century but very few have concentrated specifically on farming activities.[7] As far as historical scholarship is concerned, rural life and agrarian matters have been neglected and Fohlen stressed that geographers have made most progress in this respect even though their explicit concern for spatial relationships has been substantially different from the institutional or technological emphasis adopted by economic historians.[8]

Ever since the earliest pages of the *Annales de Géographie* in 1891 and the first regional monographs that appeared before 1914, French geographers have devoted much attention to recording the characteristics of rural life and landscapes, tracing their evolution and seeking explanations for their variation in space.[9] Very strong academic links existed until recently between geography and history and it was therefore hardly surprising that several generations of geographers saw the study of the past as the key to understanding the present.[10] The rural emphasis of so much of their work had its logic in the interrelationship between human and physical factors expressed in cultural landscapes which could be comprehended much more readily in the countryside than in the town.[11] For two generations of French geography before World War II 'cities were somehow anathema, unwelcome intrusions in the neatly ordered agricultural landscape'.[12]

As well as being overtly rural in bias a great deal of geographical work was regional in emphasis, portraying and explaining the characteristics of distinct parts of France and rarely changing scale to draw comparison with conditions in other regions or with France as a whole. This regional emphasis was both a reflection of accepted methodology, in a country where pays could still be readily distinguished, and a manifestation of expediency as academics often researched the pays, *département* or province in which they had been appointed to teach (Figure 1.1). Identical reasons help explain why some historians also cast their work into regional frameworks. Because of their academic links with geography they too paid attention to environmental resources in their study areas and to how those were perceived and used differently through time.

The net result of work by geographers and economic historians has been a vast and scholarly 'jigsaw' extending across the length and breadth of France. On closer scrutiny it becomes clear that it is not just one puzzle but several and that each is

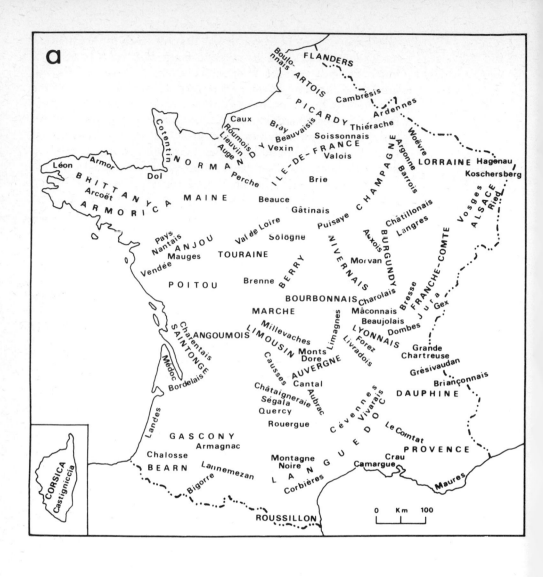

Figure 1.1: (a) Provinces and *Pays* Mentioned in the Text

6

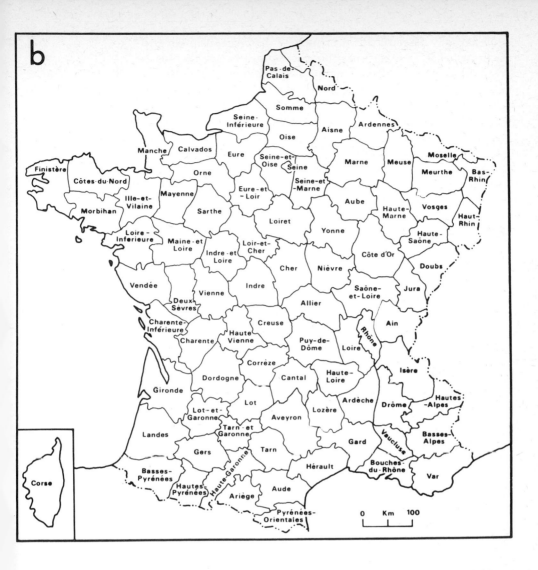

b

Pas-de-Calais

Nord

Somme

Seine-Inférieure

Oise

Aisne

Ardennes

Manche

Calvados

Eure

Seine-et-Oise

Seine

Marne

Meuse

Moselle

Meurthe

Bas-Rhin

Finistère

Orne

Seine-et-Marne

Côtes-du-Nord

Mayenne

Eure-et-Loir

Aube

Vosges

Ille-et-Vilaine

Sarthe

Haute-Marne

Haut-Rhin

Morbihan

Loiret

Yonne

Haute-Saône

Loire-Inférieure

Maine-et-Loire

Loir-et-Cher

Cher

Nièvre

Côte d'Or

Doubs

Indre-et-Loire

Vendée

Vienne

Indre

Saône-et-Loire

Jura

Deux-Sèvres

Allier

Ain

Charente-Inférieure

Creuse

Charente

Haute-Vienne

Puy-de-Dôme

Loire

Rhône

Corrèze

Isère

Dordogne

Cantal

Haute-Loire

Ardèche

Drôme

Hautes-Alpes

Gironde

Lot

Aveyron

Lozère

Lot-et-Garonne

Tarn-et-Garonne

Gard

Vaucluse

Basses-Alpes

Landes

Gers

Haute-Garonne

Tarn

Hérault

Bouches-du-Rhône

Var

Basses-Pyrénées

Hautes-Pyrénées

Ariège

Aude

Pyrénées-Orientales

0   Km   100

Corse

(b) *Départements*

7

only partially assembled. Unfortunately the pieces belonging to each set neither fit their neighbours very neatly nor the pieces from other sets. In any case many pieces are missing from each set. For the historical geographer, with his dual concern for space and time, this incompleteness of enquiry is at once tantalising and challenging. So much is known about specific themes in particular regions but when one stands back to try to encompass the nation as a whole the picture becomes much less satisfactory. Over a decade ago Landes condemned French work in regional history as 'parochial' and 'lamentably indifferent to interregional comparisons and contrasts'.[13] Exactly the same may be said of the interesting but extremely uneven *Univers de la France* series which has appeared since then and of the greater part of geographical writing.

The problem has not been resolved by various overviews of economic and social conditions for particular periods in the past or by examining critical processes, such as the modernisation of rural France.[14] No matter how stimulating such works of synthesis may be, they draw heavily on the results of published regional enquiry and usually fail to include a national frame of reference. They serve to perpetuate the spatial fragmentation of knowledge that perhaps is perceived more acutely by the foreigner who lacks profound ties to a particular pays or province and who has the urge to compare parts with the national whole. He may well feel uneasy about the representativeness of the chosen case study or the selected quotation, regardless of how elegant or colourful it may be. If he is an historical geographer he will probably be anxious to determine how the specific example fits not only into its immediate surroundings but into France as a whole. He will want guidance as to whether it may be accepted as a manifestation of the 'rule' or whether it is more likely to be an exception to that rule. It is in this latter, comparative respect that rural France in the past century may still be thought of as a *terra incognita*. Such was surely the sentiment in Laurent's mind when he called for more studies to be undertaken in the 'historical geography' of French agriculture in the nineteenth century.[15]

Cartographic manifestation of economic and social conditions in times past is central to historical geography and many historical geographers would argue that spatial representation of available information not only adds a valuable new dimension to complement national average figures, computed by historians or economists, but also permits the results of more localised enquiries to be placed in context. Preparation and comparison of a series of cartographic cross-sections allows an impression of the magnitude and direction of intervening change to be gauged. However, extreme caution is essential when comparing static pictures in time. One or even all sets of data may be pale representations of reality in times past and, in addition, the processes at work between successive cross-sections may have been infinitely more complex and even contradictory than simple comparison of cross-sections may suggest. Graphical work of this kind, together with time-series analysis of selected phenomena, has been started for the second half of the nineteenth century under the direction of Laurent. The resulting maps and diagrams form a fascinating record of agricultural responses to the slow and complicated modernisation of rural France between 1870 and 1914 following the elaboration of transport networks and the generalisation of primary education.

Earlier phases have not attracted attention and this may be for two main reasons. First, statistical enquiry in France was still in its infancy before the mid-nineteenth century; and second, the pace of economic and social change was slower than in the subsequent railway age. Such arguments are valid but only up to a point. On the one hand, a standardised agricultural enquiry was undertaken in the early years of the July Monarchy, with returns being made from every *commune* in the country and being published in 1840 and 1842 as part of the great *Statistique de la France*.[16] On the other hand, recent enquiries into French rural life in the nineteenth century and, in particular, into the so-called 'agricultural revolution' have stressed the desirability of employing quantitative as well as qualitative indicators, of moving beyond national averages toward regional and local conditions, and of placing more attention on conditions during the July Monarchy, in other words, on the eve of the railway age.[17]

During the July Monarchy France experienced the latter phase of a surge of population growth which contrasted with high birth rates plus high death rates in the early eighteenth century and with declining birth rates amid general demographic depression during the second half of the nineteenth century. Not until after 1945 was France to undergo another sustained phase of demographic vitality. Estimates of population and early censuses must be treated with caution, as must changes in national boundaries, but it would seem that the French nation living within its various contemporary boundaries grew from *c.*19,000,000 in the early eighteenth century, to *c.*21,000,000 at mid-century, *c.*26,300,000 by 1790 (including 834,600 and 124,000 from the annexation of Lorraine and Corsica), 27,350,000 in 1801 and 35,800,000 in 1851.[18] Although relatively modest when compared with the experience of other European countries such a surge of growth was truly phenomenal when seen in the context of France's very distinctive demographic history. It was all the more amazing since the French did not undergo a precocious 'industrial revolution' but remained overwhelmingly a nation of countryfolk, of peasants who survived more or less from their own resources. Basic foodstuffs for peasant farmers and for the nation as a whole were essentially home grown throughout the period. Grain imports were required in only a few very densely populated and consistently deficient areas, such as the *pays marseillais*, and of course during the terrible months that followed local or national harvest failures. This relative self-sufficiency in temperate foodstuffs made France quite unlike her British neighbour during the first half of the nineteenth century.[19]

Between the Revolution and 1850 France's agricultural resources had to be manipulated to feed, if not to satisfy fully, no less than 9 million extra mouths. Change in her agricultural systems most certainly occurred. More land was brought into permanent cultivation, new crops were grown, more livestock kept, and additional numbers of hands became available for agricultural work which, in turn, allowed agricultural intensification in some parts of the country. However there was no 'agricultural revolution' just as there was no 'industrial revolution'. As a result farming systems, the generators of food supplies, were being stretched to breaking point during the July Monarchy; poverty was widespread in some pays as well as being endemic among the lower classes. The diet of many Frenchmen was frugal in the extreme and hardship, if not actual starvation, occurred when crops failed either locally or nationally since most forms of internal transport had reached only a rudimentary stage.

Rapid population growth, barely adequate food supplies and enduring provincialism, which reflected geographical diversity and poorly developed systems of communication, were interwoven inextricably and represented fundamental 'problems' for Louis-Philippe's administrators and, indeed, for every Frenchman whose supply of daily bread was threatened. After examining each of these issues in turn the present study will attempt to reconstruct the diversity of French agricultural production at this critical phase prior to railway building which heralded the slow and intricate modernisation of rural France. Information on land use, livestock numbers and other static indicators may be used to create the frame within which the processes of agricultural production took place. Then dynamic indices, such as measures of productivity, prices and differences between supply and demand, will be analysed to demonstrate the relative efficiency of the various agricultural systems in operation during the 1830s. These will receive particular attention in the following chapters and information on calorific intake will be scrutinised since it may be argued that this represents a very telling index of agricultural efficiency.

The Keys of the Kingdom

Abundant but often highly fragmentary evidence on French agriculture during the July Monarchy is contained in the publications of agricultural and statistical societies and on the pages of journals kept by agricultural writers and official government inspectors. The records of contemporary map-makers, census-takers and tax-collectors add more detail but it is the agricultural enquiry undertaken as part of the great Statistique de la France that offers the most valuable source for constructing a nationwide view of agricultural activities in Louis-Philippe's kingdom.

Attempts had been made at various times to gather information on agricultural production from the reign of Louis XIV to the early nineteenth century but none of these represented even basic reflections of local realities. *Intendants* were largely ignorant of the provinces they administered, in spite of a number of ingenious attempts at estimating land use during the *ancien régime*. Vauban, for example, had land use surveyed for a number of sample areas in western France and results were extrapolated on the assumption that the balance of land use was uniform throughout the country.[20] The result was no more than fantasy. On the basis of his *Travels* in 1787-9 Arthur Young prepared a map of soil types and agricultural regions which was dissected by government officials and the parts weighed in order to determine the proportion of land in each category. Other estimates were even more fanciful, as for example the attempt to derive the national total of arable land from the supposed number of ploughs in 1790.

By contrast with these haphazard schemes, Napoleon I ordered that a *Statistique générale* should be collated and in 1810 each prefect was instructed to answer 33 questions within two months. In fact very few returns were made and the attempt failed completely. Rough estimates of agricultural conditions were prepared but these failed to distinguish France proper (86 départements) from the broader occupied territories, comprising a further 45 départements. In 1814 a new project for a Statistique générale was announced and an agricultural enquiry was begun.[21] This time prefects were not required to complete numerical schedules but were invited by Becquey at the Ministry of the Interior to supply observations on land use, crop production and livestock.[22] Fragments of information for only 19 départements have survived, although particularly interesting local reports are held in some archives, for example those at Rennes.[23] It soon became apparent that numerical estimates would have been preferable to impressionistic reports and beginning in 1815, prefects were required to determine the amounts and prices of various foodstuffs produced and consumed each year in their départements. The resulting figures conveyed only the broadest impression of conditions and were not drawn up for units smaller than départements.

In 1835 Louis-Philippe announced a new project for the Statistique générale that was to be masterminded by Moreau de Jonnès and in which agriculture would figure prominently.[24] In the circular of 12 July 1836 Hippolyte Passy (Minister of Commerce and Public Works) remarked that many European nations already possessed estimates of land use and agricultural production and then gave prompt instructions to each prefect as to how an enquiry was to be conducted to obtain such information for France. Prefects were to make use of every means at their disposal to obtain agricultural data from each commune. A sample questionnaire was enclosed for reprinting with the appropriate administrative heading for each département (Appendix). Mayors were to be encouraged to complete the printed schedule as accurately as possible and were to be reassured that the information was required for statistical purposes only and not for fiscal use.[25] Passy admitted that many mayors would not have such data available and instructed them to consult farmers and knowledgeable members of the community on issues such as yields and prices.[26] Information was sought for the 'average year' and it was neither required nor sufficient to supply figures for any individual year. The Minister recommended that a committee be established in each département to check, control and, if necessary, modify figures supplied by mayors. After a short delay, the prefects despatched their versions of Passy's instructions to the authorities in each of the 37,300 communes.

The first part of the questionnaire dealt with 24 types of crop (or land use) and in each case required statistical information on the land involved (in ha), the total yield (in kg, hl or m^3 as appropriate), the average price per kg, hl or m^3 (in francs and centimes), quantity used for seeding (in kg or hl) and the total amount consumed in the commune (in kg or hl). The second part of the questionnaire sought to discover the numerical importance of 17 types of livestock in each commune and the average value of each category of beast. Then information was required on the numbers of livestock (eleven categories) butchered each year, the weight of meat consumed and the average price/kg of each type of meat. Finally there was space for observations to be inserted.

In many localities the enquiry was viewed with suspicion, in spite of reassurance from the authorities that it had nothing to do with tax-gathering. Some mayors and their advisers insisted on giving literal rather than numerical replies, which were often expressed in illegible handwriting. In spite of having been introduced more than 40 years earlier the metric or 'national system', as it was called, was little understood in many communes.[27] Other difficulties included an insistence by some mayors to modify the questions to which they were supposed to be replying; the argument that such work required specially paid investigators; the fact that the total area of each commune was not known since the cadastral survey had not been completed; and the problem that local terminology for crops and land-use types did not correspond with what was shown on the questionnaire.

In order to try to overcome these difficulties a hierarchy of revision committees was established for *cantons, arrondissements* and départements with the duty of 'correcting' and standardising the form in which replies from individual communes were presented. Some prefects could not hide their frustration at having to perform so many calculations and needing to request information from foresters and tax-collectors as well as from tardy mayors. Nevertheless, the report to Louis-Philippe claimed that 'the work was greatly improved by gatherings of enlightened men who had practical understanding of farming and a knowledge of localities involved'.[28] Eventually the corrected commune returns were collected and summarised for each arrondissement. These summaries, together with the département totals, were despatched to Paris and reworked. What happened to all the commune returns is not known. There is no central repository in the *archives nationales* or in the records of the Ministry of Agriculture. A few archives contain a full collection of replies for their départements but in the majority none have survived. Additional information was sought on beer and cider in a circular of 7 December 1838 and on the value of fallow land and pastures in a circular on 23 November 1839. The net result was four quarto volumes covering the four quarters of France and containing a veritable avalanche of figures. Two pages were devoted to each département and provided a brief statistical summary, however the greater part of each volume was allocated to detailed statistics on successive topics. These were listed by département and by arrondissement for the whole of France.

The published version of the agricultural enquiry offers a number of advantages for the researcher. It is the first investigation into farming conditions to be completed for the whole country. It was undertaken approximately simultaneously in every commune and related to the average year in the 1830s. Information was expressed in metric measures rather than in the host of local systems that had been current before the Revolution and continued in use in the remote backwoods for long after 1789. The range of questions was broad involving not only land use but also yields and prices for individual crops, the relationship between production and consumption, and aspects of livestock husbandry. The results were presented in a consistent way at two spatial levels (arrondissements and départements).

As contemporaries were quick to remark, the Statistique also displays profound disadvantages because of sins both of omission and of commission. The agricultural enquiry was certainly wide-ranging but some important issues were excluded and this reduced its utility for interpretation. Most serious is the fact that no investigation was made into the number of holdings, farm sizes, tenure arrangements or the size and structure of the farming population. As an English reviewer remarked in 1848, 'on these points we are left in the dark'.[29] Information on livestock was of course included in the printed reports but it was sparse by comparison with the attention given to land use and cereals. The Statistique bears the stamp of an obsession with food supplies, in the same way as the *enquêtes* of the ancien régime and the Revolutionary years. It is not a complete *Statistique agricole* comparable with the increasingly elaborate investigations undertaken in 1852, 1862, 1882, 1892 and 1929.[30]

What was actually documented was far from being faultless and it must be admitted that the recorded figures were simply estimates of varying accuracy from commune to commune and from item to item. Nonetheless, the data were scrutinised and corrected at several levels in the administrative hierarchy and one can do no more than hope that excessive over- and under-estimates may have been averaged out as the information was synthesised for arrondissements and départements.

Useful though the arrondissement is in providing a finer level of spatial organisation than the département it is not without its shortcomings. Arrondissements are equal in administrative significance but they are far from equal in extent, with enormous but sparsely populated units in the eastern Paris Basin and Landes contrasting with small but densely populated arrondissements in Nord and Lower Normandy (Figure 1.2). The second major shortcoming of arrondissements as spatial units is that their limits normally do not correspond with those of 'natural regions'.

To a large extent the Statistique has to be taken at face value but with the very important proviso that it is an estimate and not a rigorously precise portrayal of agricultural reality. There is no comprehensive way of checking the validity of its contents since it was the first nationwide survey of French farming. Admittedly, the *ancien cadastre* contains information on broad features of land use but its purpose was different and its cat-

Figure 1.2: *Arrondissements*

egories were not precisely comparable with those of the Statistique. The cadastre was compiled over a wide span of years and was not complete in the 1830s.[31] Writing immediately after the Statistique, official agricultural inspectors tended to quote its figures in their reports rather than conducting independent enquiries. All that remains for critical comparison with the Statistique are the contents of voluminous but highly fragmentary publications in agricultural journals and later monographs. It is foolhardy to claim that the Statistique can offer more than an impression of agricultural conditions during the July Monarchy and for this reason opinions vary considerably on its utility as a historical source. Gille remarked that 'its value is very uneven and often suspect', but Augé-Laribé stressed that it had been prepared 'with care' and Musset found the figures 'sincere and coherent.'[32] For Chombart de Lauwe it was a *tour de force. . .* with results that leave something to be desired but often seem to be no less inaccurate than those of more recent enquiries.'[33]

Relatively little use has been made of the Statistique by historians and geographers. One suspects that this is largely because of the relative rarity of the volumes, which are not to be found in all *archives départementales*. Certainly the charge of dubious accuracy has not stopped researchers using other rough estimates as sources. A few major agricultural studies have employed the Statistique in some detail and many others have drawn isolated figures from it. For example, Chombart de Lauwe used département totals to compare farming in Brittany with that in the Pays de la Garonne. The same scale of approach was employed by Armengaud in his background sketch of the rural economy of eastern Aquitaine.[34] In each case one wonders why the scope of presentation was not enhanced by including information for arrondissements. Juillard, in his study of Lower Alsace, mapped data at that scale, but only Bozon, working in Vivarais, produced truly detailed maps of agricultural conditions at the commune level.[35] More recently Fel has drawn a sample of sketch maps for cereal prices but his cartographic method was extremely inferential, while Désert has plotted information on meat consumption in a more conventional and rigorous way.[36] Morineau has made use of national and département data on the production of wheat and rye in his intricate denial of the occurrence of an agricultural revolution in the eighteenth century, and he incorporated brief indications relating to green fodders, maize, potatoes and livestock in later publications.[37] His highly evocative maps of wheat and rye yields at the département scale show the long-established supremacy of Nord with remarkable clarity, for it was in Flanders that exceptional yields were being registered as early as the fourteenth century consequent upon fodder crops being alternated with cereals.

The historical geographer is prompted to ask new questions of the Statistique de la France. He will wish to know how wheat, rye and the other crops mentioned by Morineau fitted into the total pattern of land use throughout France. He will want to investigate such issues as productivity, prices and the ratio between production and consumption for each crop. He will want to examine livestock farming. He will want to obtain a more detailed spatial view than Morineau's mapping of département data allows. He may also ask how all aspects of agricultural production evolved at the regional scale prior to the 1830s. The final question is both enormous and probably unanswerable because of lack of statistical information; however the other issues may be investigated from arrondissement returns in the Statistique which are eminently suitable for cartographic display to complement the literary record of contemporary observers. Quantitative cartography is indeed central to the present study, which combines traditional and automated techniques of map construction in order to elaborate the message of words and numbers (Appendix). Before such an attempt is made to examine the Statistique systematically it is necessary to delineate the fundamental problems of rapid population growth, barely adequate food supplies and enduring provincialism which permeated agricultural activities in France on the eve of the railway age.

Notes

1. H. Sée, *Histoire économique de la France: Les temps modernes 1789-1914* (Paris, 1951), p. 91.

2. E. Juillard, A. Meynier, X. de Planhol and G. Sautter, 'Structures agraires et paysages ruraux', *AE*, vol. 17 (1957), pp. 1-188.

3. J. Marzewski, 'Quantitative history', *JCH*, vol. 3 (1968), pp. 179-91.

4. S.J.B. Clough, 'Retardative factors in French economic development in the nineteenth and twentieth centuries', *JEH*, supplement 3 (1946), pp. 91-102.

5. R.E. Cameron, 'Economic growth and stagnation in France, 1815-1914', *JMH*, vol. 30 (1958), pp. 1-13.

6. P.O. O'Brien and C. Keyder, *Economic Growth in Britain and France, 1780-1914* (London, 1978).

7. R. Laurent, *Les Vignerons de la Côte-d'Or au XIXe siècle*, 2 vols. (Dijon, 1958).

8. C. Fohlen, 'Recent research in the economic history of modern France', *JEH*, vol. 18 (1958), pp. 502-15.

9. H.D. Clout, 'The practice of historical geography in France', in H.D. Clout (ed.), *Themes in the Historical Geography of France* (London, 1977), pp. 1-19.

10. A. Meynier, *Histoire de la pensée géographique en France* (Paris, 1969).

11. E. Juillard, 'Géographie rurale française: travaux récents et tendances nouvelles', *ER*, vol. 13 (1964), pp. 46-70.

12. A. Buttimer, *Society and Milieu in the French Geographic Tradition* (Chicago, 1971), p. 118.

13. D.S. Landes, 'Recent work in the economic history of modern France', *FHS*, vol. 1, (1958), p. 76.

14. A. Jardin and A.J. Tudesq, *La France des notables, 1815-48*, 2 vols. (Paris, 1972); J. Vidalenc, *Le Peuple des campagnes: la société française de 1815 à 1848* (Paris, 1970); E. Weber, *Peasants into Frenchmen: the modernization of rural France 1870-1914* (London, 1977).

15. R. Laurent, 'Le secteur agricole', in F. Braudel and E. Labrousse (eds.), *Histoire économique et sociale de la France*, vol. 3(2) (Paris, 1976), p. 627.

16. Ministère de l'Agriculture, *Statistique de la France: Agriculture*, 4 vols. (Paris, 1840, 1842).

17. M. Morineau, 'Y-a-t-il eu une révolution agricole en France au XVIIIe siècle?', *RH*, vol. 486 (1968), pp. 299-326; M. Morineau, *Les Faux-semblants d'un démarrage économique en France au XVIIIe siècle* (Paris, 1970); W.H. Newell, 'The agricultural revolution in nineteenth century France', *JEH*, vol. 33 (1973), pp. 697-731.

18. J. Dupâquier, 'Sur la population française au XVIIe et au XVIIIe siècle', *RH*, vol. 239 (1968), pp. 43-79.

19. R. Peet, 'Influences of the British market on agriculture and related economic development in Europe before 1860', *TIBG*, vol. 56 (1972), pp. 1-20.

20. A. Moreau de Jonnès, 'Statistique agricole de la France orientale', *Le Cultivateur*, vol. 16 (1840), pp. 491-502, 549-59.

21. O. Festy, 'Les enquêtes agricoles en France de 1800 à 1815', *RHES*, vol. 34 (1956), pp. 43-59; O. Festy, 'Les progrès de l'agriculture française durant le Premier Empire', *RHES*, vol. 35 (1957), pp. 266-92.

22. J. Chombart de Lauwe, *Bretagne et Pays de la Garonne: évolution agricole comparée depuis un siècle* (Paris, 1946), p. 24.

23. AD Ille-et-Vilaine, 32 Ma 1. *Statistiques*: Agriculture, an IX-1815.

24. E. Levasseur, *Histoire de la population française*, 3 vols. (Paris, 1889, 1891, 1892) vol. 1, p. 60.

25. A. Gouin, *Rapport au Roi sur le quatrième volume de la Statistique de la France: partie agricole* (Paris, 1840)

26. Ministère de l'Agriculture, *Statistique de la France*, vol. 1, p. xvi.

27. R.D. Anderson, *Education in France, 1848-1870* (Oxford, 1975), p. 5.

28. Ministère de l'Agriculture, *Statistique de la France*, vol. 1, p. xvii.

29. Anon., *Review of the Agricultural Statistics of France* (London, 1848), p. 6.

30. G. Garrier, 'Les enquêtes agricoles du XIXe siècle: une source contestée', *CH*, vol. 241 (1967), pp. 105-13.

31. H.D. Clout and K. Sutton, 'The cadastre as a source for French rural studies', *AH*, vol. 43 (1969), pp. 215-23.

32. B. Gille, *Les Sources statistiques de l'histoire de la France des enquêtes du XVIIe siècle à 1870* (Paris, 1964), p. 200; M. Augé-Laribé, 'Les statistiques agricoles', *AG*, vols. 53-4 (1945), p. 81; L. Musset, 'Observations sur l'ancien assolement biennal du Roumois et du Lieuvin', *AN*, vol. 2 (1952), p. 147.

33. Chombart de Lauwe, *Bretagne*, p. 29.

34. A. Armengaud, *Les Populations de l'Est aquitain* (Paris, 1961).

35. E. Juillard, *La Vie rurale dans la plaine de Basse-Alsace* (Paris, 1953), p. 262; P. Bozon, *La Vie rurale en Vivarais* (Paris, 1961), p. 90.

36. A. Fel, 'Géographie des productions et des prix agricoles'. Unpublished abstract of paper presented at the Conference on Rural Landscape and Settlement Evolution in Europe, Warsaw, September 1975; G. Désert, 'Viande et poisson dans l'alimentation des Français au milieu du XIXe siècle', *AESC*, vol. 30 (1975), pp. 519-36.

37. M. Morineau, 'Révolution agricole; Faux-semblants; Révolution agricole, révolution alimentaire, révolution démographique', *ADH* (1974), pp. 335-71.

Quantity

The century between 1750 and 1850 formed a time of demographic 'explosion' in France. Figures throughout the eighteenth century are open to debate but Vauban's estimate for the early years gave a total of 19,000,000 within the national boundaries of the time and Henry has revised the results to c. 22,000,000 for the territory of modern France.[1] Toward the end of the ancien régime 960,000 extra souls were brought into the French realm by the annexation of Lorraine and Corsica. In the new century the national population rose from 27,300,000 in 1801, when the first official census was conducted, to 33,540,000 in 1836, which coincided quite well with the agricultural enquiry of the Statistique. To achieve a numerical increase of such dimensions had taken the full span of the eighteenth century and was to take no less than 75 years after 1836, with the national total reaching 39,610,000 on the eve of World War I.

How had the French population evolved during those critical three and a half decades; how was it located across the face of the country; and what were the implications of this distribution for the vexed problem of ensuring an adequate supply of foodstuffs? Answers to these questions may be sought at the département and arrondissement scales from successive censuses which recorded population totals but gave no information on natural increase, migration, employment or urbanisation. There is, for example, no way of determining how many migrants returned to France in the early years of the Empire or left at the beginning of the Restoration. Some contemporaries talked of more than a million in each case.[2] Census-taking is always fraught with difficulties and nineteenth-century France was no exception. However it is likely that the hazards of interpretation may be kept to a minimum by working at a fairly coarse spatial scale and by ignoring censuses that are recognised as unreliable. For example, the census of 1801 has been described as 'an excellent point of departure', unlike that for 1806, which contained many exaggerations, and the so-called censuses of 1811 and 1816 which were only projections from earlier figures.[3] Census-taking became more rigorous in 1821 but the first nominal rolls were not prepared until 1836. Only figures for 1801, 1821, 1826, 1831 and 1836 have been retained for scrutiny in the present chapter. Boundary changes have been taken into consideration wherever possible, for example where new arrondissements or new départements (e.g. Tarn-et-Garonne, 1808) were created, but there is no way of adjusting for minor modifications.

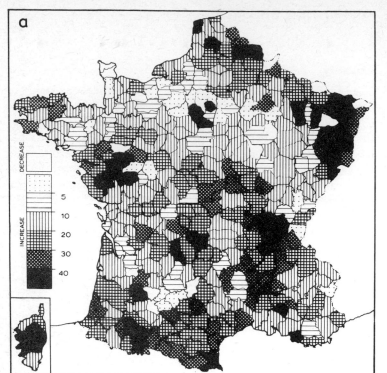

Figure 2.1:
Population Change
(a) Population Change,
1801-36 (%)
(b) Losses for
Intercensal Periods,
1821-36

a

DECREASE

INCREASE

5
10
20
30
40

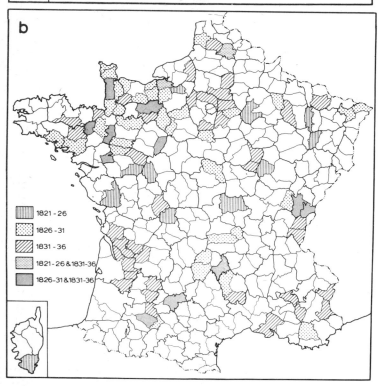

b

1821 - 26
1826 - 31
1831 - 36
1821 - 26 & 1831-36
1826 - 31 & 1831-36

(c) Towns With More
Than 10,000
Inhabitants, 1836
(d) Rural Population
Change, 1801-36 (%)

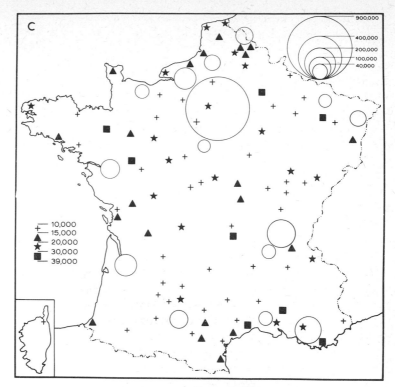

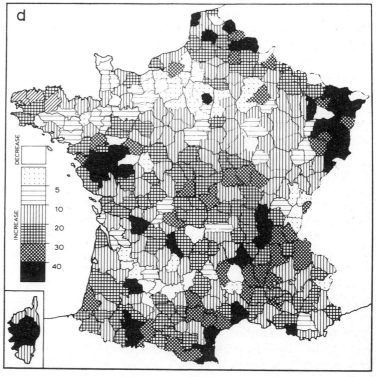

Demographic growth between 1801 and 1836 was far from constant, with warfare at the end of the Empire, typhus in eastern France in 1813, famine in 1817, food shortages in 1830 and 1831, and cholera in 1832 and 1834 contributing to distortions on the graph. Some years were characterised by high birth rates, such as 1814 following the large number of marriages contracted in 1813 in order to avoid conscription, and stood in contrast with 'hollow years' such as 1812 and 1813. The average annual increase in population between 1821 and 1846 was some 330,000, roughly double the figure for the previous 20 years. The first four decades of the nineteenth century were truly remarkable in the context of French demography and in some regions fragile agricultural systems were subjected to severe strain. Food shortages were typical in *mauvais pays* and in poor years, such as 1846 and 1847, crisis conditions occurred widely. Such disasters broke the continuity of population dynamism, unleashed an extensive rural exodus and heralded the slow rates of demographic growth that were to characterise France for the remainder of the nineteenth century. When the long span of 1815 to 1865 is examined, the population of France grew by a mere 0.46 per cent per annum, only a third as fast as England and Wales, Norway, Prussia or Saxony.

None the less, between 1801 and 1836 the French total increased by 22.6 per cent with growth being recorded in every département.[4] Rates of change were extremely uneven, ranging from more than double the national average in Seine (up 75 per cent), Rhône (up 61 per cent) and Pyrénées-Orientales (up 48 per cent) to less than half the average in a dozen départements in the Paris Basin, Aquitaine and the Jura. Seventeen widely dispersed arrondissements displayed population losses when 1836 figures are compared with those for 1801 (Figure 2.1a). These early losses may have been more apparent than real, since there may have been errors in enumeration, more probably for 1801 than for 1836. In addition, there may have been small boundary changes between the two dates and the figures for 1801 may have been inflated by armed forces in some of the frontier and coastal areas in which 'losses' were recorded subsequently. Net out-migration might have outweighed (or almost outweighed) natural increase in some départements. This type of argument would seem quite plausible for explaining the limited growth in some arrondissements near Paris, since sample studies for the eighteenth century showed that the capital was exercising a blanketing effect on migratory behaviour for up to 270 km from the city.[5] Similarly a study relating to 1833 showed the importance of départements in central, northern and eastern parts of the Paris Basin in supplying migrants to the capital.[6] By 1835 concern was being expressed in Brie that there was not

22

enough local labour to bring in the wheat harvest and such may have been the case in other areas close to the capital.[7]

Circumstances may well have been peculiar in many parts of France at the opening of the nineteenth century, with for example eastern départements forming the theatre of operations for troop movements during the Empire, but by the third decade conditions were more settled and censuses were conducted on a more reliable basis.[8] It is clear that some arrondissements were losing population well before the food crisis of the late 1840s. Between 1821 and 1826 numbers declined in only 13 arrondissements but rose to 45 between 1826 and 1831 and were to increase to 52 between 1831 and 1836 (Figure 2.1b). The latter total included eleven arrondissements which had recorded losses between 1826 and 1831. However in eleven arrondissements the losses between 1821 and 1826 were not repeated in the next 10 years and for 36 arrondissements the decline recorded between 1826 and 1831 was not repeated during the first half of the 1830s. In such areas population loss was only a transitory phenomenon. It was, none the less, an indication of the inability of local economic systems to support their population and heralded the more pronounced depopulation that was to occur later in the century.

Losses in the first third of the nineteenth century were significant but they were relatively few and far between. No fewer than 346 arrondissements contained more population in 1836 than in 1801. Very high rates of growth (40 per cent or more) affected not only arrondissements containing major cities but also a wide range of more obviously 'rural' environments in Alsace, Lorraine, the Pyrenees, Vendée and Massif Central; and increases of between 30 and 40 per cent were widespread in central, southern and north-eastern France. Twenty arrondissements grew by 50 per cent or more and in five the population more than doubled. These were Saint-Denis (up 175 per cent) and Sceaux (up 101 per cent) on the margins of Paris; Bourbon-Vendée (up 121 per cent) where internal colonisation had operated; Brive (up 139 per cent) and Montbéliard (up 111 per cent). Only in the inner Paris Basin and the basin of the Garonne were there sizeable clusters of contiguous arrondissements with very low rates of growth.

An official definition of urbanisation was not introduced in France until 1846 when communes containing nucleations of more than 2,000 inhabitants apiece were deemed to be 'urban'. In that year 24.4 per cent of the national population were town dwellers. However, the statistics for 1801 and 1836 do give numbers living in the chef-lieu, which was normally the most populous settlement of each arrondissement. Many chefs-lieux were quite small and only 125 contained more than 10,000 apiece which together accounted for 3,569,976 in 1836 (Figure 2.1c). Large towns of more than 15,000 inhabitants were found especially in northern France and along the Mediterranean littoral. Paris, with 909,100 inhabitants in 1836, was six times the size of Lyons (150,810) or Marseilles (146,240). Six other cities contained over 50,000 residents apiece (Bordeaux 98,705, Rouen 92,083, Toulouse 77,372, Nantes 75,895, Lille 72,005, Strasbourg 57,855). These nine cities formed the nation's major demand centres but contained only five per cent of the total population. Even when the remaining 116 towns with between 10,000 and 50,000 inhabitants are included the proportion rises only to 10.6 per cent and to 14.8 per cent when the 4,951,684 residents of all chefs-lieux are considered. Thirty-five years earlier the same 363 chefs-lieux had housed 3,854,202 people (14.1 per cent of the total) and the 28.5 per cent increase over the intervening years was not much greater than the growth rate for the whole nation (22.6 per cent).

In spite of the absence of official definitions the overwhelming rurality of France in the 1830s was unquestionable. Communes with fewer than 3,000 inhabitants made up 96.8 per cent of the country's 37,352 first-order administrative units and housed 75.4 per cent of her people.

Rural France, that is France beyond the chefs-lieux d'arrondissement, increased its population by 21.7 per cent between 1801 and 1836, accounting for no less than 82.3 per cent of population growth (Figure 2.1d). As one would expect for such a strongly rural society the pattern of population change in the countryside approximated closely to that for complete arrondissements. The rural parts of 19 arrondissements grew by more than 50 per cent and in seven of these the population more than doubled. Four functioned as reception areas on the margins of the nation's largest cities (Saint-Denis up 182 per cent and Sceaux up 103 per cent; Marseilles up 121 per cent; Lyons up 100 per cent) and there were three isolated examples (Brive up 151 per cent, Montbéliard up 121 per cent, Bourbon-Vendée up 115 per cent).

France supported 63.6 persons/km² in 1836 and in 15 arrondissements densities were more than twice as great (Figure 2.2). These included the major urban centres and their satellites, two fertile and intensively cultivated Flemish arrondissements (Hazebrouck 153/km², Dunkirk 134/km²) but also the very different arrondissement of Ambert (134/km²) high in the eastern Massif Central. These population peaks were set in five zones which supported more than 100 persons/km². The first extended along the Channel coast from northern Brittany through Lower Normandy to Le Havre, Rouen and the Pays de Caux. Nord plus Pas-de-Calais and the arrondissement of Saint-

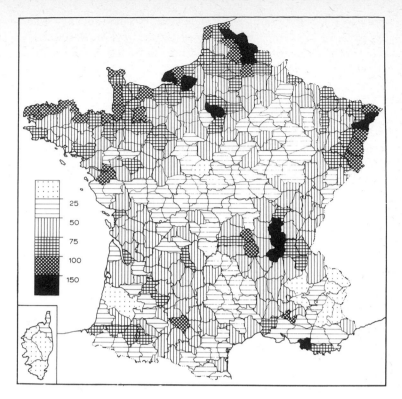

Quentin (Aisne), and Bas- and Haut-Rhin formed the next two
zones. By contrast with these quite extensive areas, greater Paris
occupied only four arrondissements but three of these contained
the highest densities of all (Paris 2,655/km², Saint-Denis
544/km², Sceaux 367/km²). Lyons formed the centre of the fifth
densely peopled zone, which embraced the arrondissements of
Saint-Etienne and Villefranche, and with which the detached
arrondissements of Ambert and Thiers (Puy-de-Dôme) might be
associated, since Lyons and Saint-Etienne commanded important
migration fields and regions of industrial out-working in these
eastern sections of the Massif Central. Only four other arrondissements contained over 100 inhabitants/km², namely the
relatively free-standing cities of Marseilles (274/km²), Avignon
(140/km²), Nantes (117/km²) and Toulouse (101/km²). Environments as diverse as sections of the Paris Basin, the Massif
Central, the basin of Aquitaine and central Brittany supported
between 50 and 75 people/km², while the Alps, Sologne, Landes
and Corsica formed truly empty lands. When the population of
chefs-lieux is excluded the rural density falls to 54.0/km². The
general pattern remains similar to that presented already but

only in the Seine (Saint-Denis 498/km², Sceaux 360/km²) and Nord (Lille 271/km², Valenciennes 176/km², Douai 160/km², Cambrai 156/km²) were rural densities recorded that were three times or more the national average.

Quality

It is not possible to determine the structure of employment in the 1830s since the first census to pose questions on this issue was not undertaken until 1851. In any case any attempt to define agricultural, industrial and service employment with any precision at this stage of economic development must be open to debate since many workers undertook varying types of activity at different times of the year and sometimes incorporated seasonal out-migration with farm work. However it is possible to re-work three sets of information published by D'Angeville in 1833 in order to gain a rough measure of spatial differences in employment during the early July Monarchy.[9] Fifty-three per cent of army recruits between 1825 and 1833 were recorded as *agriculteurs* but there was a clear contrast between low percentages in northern and north-eastern France and high proportions in central and western France where they made up over two-thirds of recruited men. In Ardèche and Haute-Loire 76 per cent of recruits were agriculteurs, by contrast with 11 per cent in Seine-Inférieure and 18 per cent in Somme. According to D'Angeville 22 per cent of recruits were *industriels* between 1825 and 1829. They were particularly numerous in seven départements in the Paris Basin and on its margins, with more than half the recruits in Seine, Seine-Inférieure and Somme being industriels, compared with only eight per cent in Haute-Loire, Hautes-Alpes and Morbihan.

Craft industries remained important in many parts of the French countryside in the 1830s although perhaps less so than in the late eighteenth century. Rural crafts played a vital role in supporting high population densities, as in the harsh surroundings of Ambert which manufactured paper, cloth, ribbons and lace, using linen thread from Nord and Holland.[10] Similarly the high densities in rural Normandy and Maine were supported by textile-making, tanning, iron-making and other non-agricultural pursuits. Doublet de Boisthibault went so far as to describe the départements of Eure, Orne, Calvados and Manche as 'the most productive and most industrialised in France'.[11] His view was exaggerated but the early onset of rural depopulation in these areas may be attributed in part to the ruin of linen-making and other craft industries in the late-eighteenth and early-nineteenth centuries.

In 1819 Chaptal claimed that misery had been banished from the French countryside and 'well being had been born of the ready exchange of agricultural goods'; countryfolk enjoyed 'an abundance of food, had clean homes and simple, decent clothing'.[12] His views may have reflected conditions in prosperous countrysides but were certainly over-optimistic for the *mauvais pays* and possibly for the greater part of France. Two decades later the agronomist Lullin de Châteauvieux argued that the nation's agricultural production had increased by 36 per cent between 1789 and the late 1830s and that as a result, and speaking in average terms, the 33,500,000 inhabitants of France enjoyed better diets, clothing and housing than their 25,000,000 predecessors who had been described by Arthur Young.[13] Wealth and well-being had not, of course, increased evenly across the face of France and Adolphe Blanqui could claim at the mid-century that the contrast in living standards between town dwellers and countryfolk was 'the single economic fact that deserved the greatest attention'.[14] He argued that there were 'two different peoples living such different types of life in the same country that they seem complete strangers to each other'. But the contrast between town and country was not really so clear cut nor were some inhabitants of towns and cities in the July Monarchy short of their share of misery. More fundamentally, there were great variations in the quality of life not only between pays but also between different strata of rural society.

Nineteenth-century observers attributed such variations to differences in the nature of the agricultural environment and also to variations in the mentality of the peasantry. *Mauvais pays* were characterised by widespread human poverty. For example, the Sologne was 'the most disgraceful part of the national territory, after the Landes of Gascony, the soil is drowned in winter and burnt in summer: men, animals and plants, all assume a wretched appearance'.[15] Things were so bad that in the midst of these 'uncultivated steppes, only five or six hours from Paris, one might well be back in the middle ages'.[16] Children succumbed most readily to fever in the unhealthy climate of the Sologne and the neighbouring Gâtinais and 'young people were consumed before reaching an age to be useful'.[17] The material condition of peasants in the mountains of central France was 'particularly bad, because of the type of farming' whilst 'words could not convey a just impression of the harsh and severe lives led by inhabitants' of the Alps.[18] By contrast, life was held to be comfortable in more fertile pays. It was claimed that good farming in Nord supported a rural population which enjoyed good food, were healthy and clean of habits, loved work and observed their religious duties.[19] In Normandy 'servants were fed as well as their masters', while the peasants' houses in

Gironde 'could pass for palaces when compared with the miserable shelters of cultivators in Lozère, the Alps or Brittany, most of whom still use the same rudimentary implements that their ancestors used in the fifteenth century'.[20]

Even in rich countrysides poverty was not hard to find. The granary of Beauce was 'one of the most productive' farmlands in France, supplying wheat to Paris, and yet it also displayed one of the sharpest contrasts between rich and poor in the whole country, with day labourers being fed black bread, cheese and vegetables and drinking only water.[21] The vine-growers of Tonnerois and the cereal farmers of Sens enjoyed a degree of comfort that placed neighbouring smallholders and sharecroppers in 'another world', but the general affluence of peasants in the Pays d'Auge was marred by alcoholism which 'gave young people a yellowish look, and imparted ugliness and precocious old age'.[22] Even Alsace, which had been transformed into 'a vast garden' by its progressive farmers, displayed poverty, begging and out-migration which were 'no less worrying than in the most disgraceful provinces' of France.[23]

Blind acceptance of tradition was widely blamed for rural poverty, as in the Limagnes where 'the fruitful soil might be covered with the richest harvests, if the prejudices of the peasantry were not opposed to every improvement', and in the *bocage* of Vendée where ignorance and prejudice were mainly responsible for agricultural backwardness.[24] By contrast in Moselle 'the peasant adopts willingly every improvement, raises a great variety of products and multiplies different kinds of fruits'.[25] In a broad review of agricultural productivity, Madame Romieu concluded 'the southerner counts on the sun, the northerner counts on himself'.[26] Bretons and Normans, although neighbours, were held to be quite different in their outlook. According to Souvestre, the Breton lived for 'black bread every day, drunkenness on Sunday and a straw bed to die on when he was sixty, such was the peasant's present and future which was accepted with resignation. Poverty was taken for granted as an hereditary and incurable disease'.[27] Bretons were 'loyal but savage, sleeping inside *lits-clos* in wretched cottages where farmyard animals lived under the same roofs as their masters', while Normans were 'shrewd and enlightened' and were well housed in spacious, well-ventilated and impeccably clean farmhouses.[28]

Evocative but oversimplified impressions such as these abound in agricultural journals and in special surveys of material conditions in the countryside. That undertaken by Adolphe Blanqui in 1849 and 1850 at the behest of the *Académie des Sciences Morales et Politiques* provides a powerful and sympathetic view of the peasantry at the end of the July Monarchy and was prepared on the basis of personal observations of peasants 'at work, in church, in school and in the *mairie*' in every département.[29] Toward the end of the century Henri Baudrillart attempted an extensive but unfinished review of the agricultural population which set nineteenth-century conditions into their historical context.[30] Frédéric Le Play's monographs of family life, with their careful budgets and inventories of possessions, provide a much more objective body of information relating to the second half of the century.[31] These and other sources, such as military surveys prior to the compilation of the *carte d'état major*, have been skilfully assembled and interpreted by Vidalenc and Weber to convey a much grimmer picture of peasant life than an examination of France's many and varied agricultural resources might suggest.[32]

Vidalenc could report without exaggeration that the country was still very close to a stage of under-development during the July Monarchy and displayed conditions similar to those that were to be witnessed in the French colonies a century later. Poverty was the dominant feature of the countryside and the life of the peasantry was generally difficult and wretched. Food shortages and destitution were still very real problems as crisis years showed all too clearly. Blanqui argued that the most general characteristic of the peasantry was distress and general insufficiency of the means of satisfying the first necessities of life. Yet the peasants were the most numerous group of tax-payers and a force to be reckoned with in the political life of the nation. In spite of their hardships, Blanqui could argue that they enjoyed some advantages, being 'more robust, healthy, moral, independent and happy than urban populations'.[33]

By contrast with such rich but fragmented literary description, surprisingly little numerical information about those who worked the land is available. Statistics on the height, health, literacy and employment of conscripts have been analysed with varying degrees of sophistication to demonstrate a basic contrast in French society to either side of a line from Granville to Lake Geneva. Northerners were generally taller, healthier, more literate and less likely to work the land than their compatriots in the centre and south of France.[34] Even so, there were important local contrasts in physical characteristics with, for example, powerful, well-fed peasant 'giants' in the wheatlands of the Causses contrasting with their smaller and less vigorous neighbours in the Ségalas.[35] Likewise in Brittany 'there was little in common between the solid sailors and girls along the north coast and the miserable sufferers from rickets in the interior'; whilst Solognats seemed to be simply 'a bastardised version of the human race'.[36]

From a systematic or scientific point of view very little was known in the 1830s about the French population beyond its crude numbers. Variations in living standard were alluded to by observers but were not expressed in quantitative terms. However, the Statistique contains an interesting array of information on food consumption that will be examined in Chapter 11. Even employment was not recorded in censuses until after 1850 and there is no information on the size of the agricultural workforce in the early years of the July Monarchy. Criteria by which the workforce was defined and estimates of its size vary considerably: Schnitzler advanced at least 18 million whilst Blanqui favoured an agricultural population of 25 million in 1850 and the Statistique agricole suggested 19 to 20 million in 1862.[37]

It is clear that landownership had become more fragmented after the Revolution with the number of entries of property ownership in the great cadastral survey increasing from 10,083,731 in 1815 to 10,998,730 in 1839 and 12,393,366 at mid-century.[38] Unfortunately that source is not as meaningful as it might first appear since a landowner with property in two or more communes would be recorded as a separate entry in each. Blanqui (1851) believed that out of a total agricultural population of 25 million only 2 million were landless, with 14 million small proprietors together owning a quarter of the land surface and a further 9 million owning the remainder.[39] Only after 1850 did information on the structure of farm ownership and systems of land occupation become available.[40] Undoubtedly much detail had changed in the years following the mid-1830s but a brief sketch will serve to convey the basic characteristics of French farmers and their holdings.

In 1862 about a tenth of the national population was made up of c. 3,800,000 cultivators who owned all or part of the land that they worked. Almost all of them (1,812,500) were exclusively owner-occupiers, who were entirely involved in managing their own plots, and the remainder (1,987,000) leased additional land as tenants or sharecroppers or else hired themselves out as day labourers to other farmers. A further 588,000 tenant farmers and sharecroppers possessed no land of their own and were dependent on landlords for the holdings that they worked. Almost 3,000,000 farm servants of both sexes, labourers and farm managers made up the rest of the so-called 'working' agricultural population of c. 7,363,000 souls. But farming was a way of life for an estimated 11,647,000 family members who helped work the land in one way or another. The total agricultural population was at least 19,000,000 out of a national total of

37,386,000 recorded in the 1861 census (when Nice and Savoie were included).

According to the official figures, in their unmodified form, 3,799,759 cultivators (72.3 per cent) owned all or part of the land that they worked, while a further 1,457,314 cultivators (27.7 per cent) were landless. Owner occupation was most pronounced in southern and eastern départements, and exceeded 90 per cent of the total in Basses- and Hautes-Alpes and Puy-de-Dôme. Rates of owner occupation were low in the north-west, most of the Paris Basin and in parts of the Massif Central, with only one-third of farmers in Mayenne and Seine-Inférieure being owner-occupiers. A fundamental north/south contrast in tenurial types was clearly in evidence in the middle years of the nineteenth century. In fact, it may be appropriate to reduce the official figures for *cultivateurs non-propriétaires* by excluding 869,254 *journaliers* who would better be described as wage-earners and have been designated as such in Table 2.1. The distinction between farmers who worked their own land (52.4 per cent) and those who worked in various capacities on other people's land provides a roughly similar pattern to that already discussed.

Table 2.1: Working Agrarian Population, 1862

Owner-cultivators:	farming only their own land	1,812,573
	farming their own land but also	
	working as tenant farmers	648,836
	working as sharecroppers	203,860
	working as day-labourers	1,134,490
		3,799,759
Non-owners:	tenant farmers	386,533
	sharecroppers	201,527
		588,060
Wage-earners:	managers	10,215
	day labourers	869,254
	farm servants, cowmen,	
	shepherds	2,095,777
		2,975,246
Total working agrarian population		7,363,065

According to the Statistique agricole owner-occupation 'might be considered the most fruitful' form of land holding, since the owner had 'an obvious interest in improving his property which he cultivated as a good father'.[41] Tenants on short leases, it was argued, rarely undertook any improvements and tended to exhaust the soil during the final seasons of their tenancy unless they were carefully watched. Such problems tended to diminish as the length of lease increased and hence the landlord would be assured of a reasonable return. Finally, sharecroppers 'offered all the problems of ignorance linked to poverty. Their apathy and hostility to the spirit of progress were the result of their precarious existence' since they rarely had legal entitlement to their holdings for more than one year.[42] Living day by day was enough for them; because of their ignorance of French and reliance on *patois* they were completely deprived of even the simplest notions in agriculture as well as being unable to write and add up.

Over one million cultivators leased land in 1862, with almost 650,000 farming leased plots in addition to their own property. Sharecroppers were substantially fewer in number (*c.* 405,000) with half of them also farming land of their own. Sharecropping was on the decline, being replaced by tenancy arrangements in many instances. For every seven tenant-farmers there were just under three sharecroppers in 1862. This form of land occupation was found predominantly south of the Loire, with the major concentration being in Aquitaine. By contrast with 1,035, 369 tenants, only 568,688 tenanted holdings (*fermes*) were listed in the agricultural census. Ambiguities of definition and exclusion of tenanted plots (as opposed to full holdings) contributed to this difference. But taking the official, low figure only 17.6 per cent of holdings were occupied by tenants and these were much more significant in the northern third of France than in other parts of the country. Fifty-eight per cent of leases lasted for nine years or more and these long-lease farms were particularly numerous in the Paris Basin and in northern France which included many of the most prosperous parts of the French countryside. By contrast 25 per cent lasted only 6 years and the remaining 17 per cent three years or less. As the co-ordinator of the Statistique agricole remarked, a long lease is one of the best conditions for good cultivation, while short leases of three or six years explained why agriculture made so little progress.[43]

The mean size of holding was 10.5 ha in 1862, with 10 ha being recognised as an approximate division between *petite* and *grande culture*. In fact the terms were often used without any precise qualification and there were important local differences in what might be defined as large and small farms. Three quarters of all farms fell below the 10 ha threshold, with southern départements plus Seine and Seine-et-Oise containing above average proportions. Tiny farms of less than 5 ha made up 56 per cent of the total, exceeding that proportion in 37 départements located in northern, north-western and Mediterranean France, to either side of the Gironde and in a stretch of territory from Paris to the Alps. In Bas-Rhin no fewer than 83.7 per cent of holdings were smaller than 5 ha apiece. Land was sold by the *are* rather than the *hectare* and ownership of some meadows was so fragmented and parcels so narrow that farmers were scarcely able to use a scythe without cutting their neighbours' hay. [44] Plots that were only two paces wide were not rare. These départements contained not only the largest urban centres, the most densely settled valleys and the most fertile soils but also many of the vinegrowing districts of France.

At the département scale at least miniature holdings were below average importance throughout Armorica, the outer Paris Basin, much of Aquitaine and sections of the Massif Central, except densely settled pays such as the fertile Limagnes in Puy-de-Dôme and the hillsides of Ardèche from which stones had been painstakingly cleared to form terraces. In Velay and the Limagnes holdings were divided to infinity with each heir taking a portion of each parcel and were cultivated by the spade so that virtually no land was wasted. However most of these regions contained important proportions of slightly larger farms, ranging in size between 5 and 10 ha. More extensive holdings of between 10 and 40 ha were widespread in the eastern Paris Basin, Armorica and its marchlands, Poitou, Limousin and parts of Aquitaine; while large farms of more than 40 ha were particularly important in the Landes, across the plains of Champagne, and in a triangle from Beauce to Bourbonnais and westward to Vendée. A wide variety of soil conditions was involved in these areas of large farms, ranging from rich arable soils in Caux and Beauce, through the inferior plain of Poitou, to the poor environments of the Landes and the Massif Central.

In the 1830s France was predominantly a nation of country dwellers and small landowners who worked the land they owned. In Labrousse's words 'it was a world that scraped by more or less, and sometimes even succeeded in setting something aside'. [45] The typical settlement was the village, with large urban centres being few and far between. With over 900,000 inhabitants Paris was truly a monster, as contemporaries observed, but even so cultivators farmed inside the line of the city walls. With such a reduced scale of settlement being the norm it may be reasonably assumed that most countryfolk were food producers and that most towns and villages derived the greater part of their foodstuffs from their immediate surroundings. Such was not, of course, the case for large settlements or for the supply of special commodities which would have to be transported from other environments. Spatial variations in the quality of internal communications by road and water (and to a lesser extent of coastal shipping) played a vital role in delimiting areas in which the peasantry might respond to urban demands by developing commercial forms of production and others in which there were no local alternatives to provincial backwardness until the railway and associated feeder roads offered opportunities for agricultural specialisation to those who were ready and able to respond. This was certainly not yet the case in the early years of the July Monarchy.

Notes

1. L.Henry, 'The population of France in the eighteenth century', in D.V. Glass and D.E.C. Eversley (eds.), *Population in History* (London, 1965), pp. 434-56; J. Dupâquier, 'Sur la population française au XVIIe et au XVIIIe siècle', *RH*, vol. 239 (1968), pp. 43-79.
2. C. Pouthas, *La Population française pendant la première moitié du XIXe siècle* (Paris, 1956), p.18.
3. Ibid. p.17.
4. Ministère des Travaux Publics, *Statistique de la France, Territoire et Population* (Paris, 1837), pp. 200-12, 267-83; E. Van de Walle, 'France', in R. Lee (ed.), *European Demography and Economic Growth* (London, 1979), pp. 123-43.
5. L. Henry and D. Courgeau, 'Deux analyses de l'immigration à Paris au XVIIIe siècle', *P*, vol. 26 (1971), pp. 1073-92.
6. L. Chevalier, *La Formation de la population parisienne au XIXe siècle* (Paris, 1950), p. 58.
7. AD Seine-et-Marne, M 7323. 'Situation et progrès de l'agriculture, 1835'.
8. J. Vidalenc, *Le Peuple des campagnes: la société française de 1815 à 1848* (Paris, 1970), p. 109.
9. A. D'Angeville, *Essai sur la statistique de la population française* (Paris, 1836); E. Le Roy Ladurie (ed.), *Essai sur la statistique* (Paris, 1969).
10. Anon., 'Statistique industrielle de l'arrondissement d'Ambert', *ASLIA*, vol. 1 (1828), pp. 209-16.
11. J. Doublet de Boisthibault, *La France: description géographique, statistique et topographique: Eure et Loir* (Paris, 1836), p. 154.
12. J.A. Chaptal, *De l'Industrie française* (2 vols., Paris, 1819), vol. 1, p. 153.
13. F. Lullin de Châteauvieux, *Voyages agronomiques en France* (Paris, 1843), p. xxi.
14. A. Blanqui, 'Les populations rurales de la France en 1850', *APAPER*, vol. 24 (1851), p. 189.
15. G. Léonce de Lavergne, *Economie rurale de la France depuis 1789*, 2nd edn. (Paris, 1861), p. 60.
16. Ibid., p. 355.
17. M.A. Puvis, 'Observations sur la nature de sol et de la culture du plateau du Gâtinais et de la Sologne', *AAF*, vol. 11 (1833), p. 210.
18. Léonce de Lavergne, *Economie rurale*, p. 402; Blanqui, 'Les populations', p. 203.
19. Inspecteurs de l'Agriculture, *Agriculture française: Nord*, (Paris, 1843), p. 25.
20. Léonce de Lavergne, *Economie rurale*, p. 91; Blanqui, 'Les populations', p. 202.
21. Malte-Brun, *Universal Geography* (Boston, 1833), vol. 8, p. 316; Vidalenc, *Le Peuple*, p. 20.
22. Vidalenc, *Le Peuple*, p. 90, 172.
23. Blanqui, 'Les populations', p. 214.
24. Malte-Brun, *Universal Geography*, p. 355; J.A. Cavoleau, 'Description du département de la Vendée', *AAF*, vol. 3 (1818), pp. 364-82.
25. Malte-Brun, *Universal Geography*, p. 401.
26. Mme Romieu, *Des Paysans et de l'agriculture en France au XIXe siècle* (Paris, 1865), p. 848.
27. Cited in H. Baudrillart, *Les Populations agricoles de la France: Normandie et Bretagne* (Paris, 1885), p. 445.

28. Blanqui, 'Les populations', p. 212.

29. Ibid. p.216.

30. H. Baudrillart, *Populations agricoles* (3 vols., Paris, 1885, 1888, 1893).

31. F. Le Play, *Les Ouvriers européens* (6 vols., Paris, 1879).

32. Vidalenc, *Le Peuple*; J. Vidalenc, 'La Corse vue par des officiers sous la monarchie constitutionnelle, 1822-47', *ACSS* (1963), pp. 65-82; E. Weber, *Peasants into Frenchmen: the modernization of rural France 1870-1914* (London, 1977).

33. Blanqui, 'Les populations', p. 218.

34. D'Angeville, *Essai sur la statistique*; E. Levasseur, *Histoire de la population française* (3 vols., Paris, 1889, 1891, 1892); J.P. Aron *et al.*, 'Anthropologie de la jeunesse masculine en France au niveau d'une cartographie cantonale, 1819-30', *AESC*, vol. 31 (1976), pp. 700-60.

35. H. Baudrillart, *Populations agricoles: Midi* (Paris, 1893), p. 450.

36. A. Jardin and A.J. Tudesq, *La France des notables, 1815-48* (2 vols., Paris, 1972), vol. 2 p. 12, 171.

37. J. Schnitzler, *Statistique générale méthodique et complète de la France* (4 vols., Paris, 1846); Blanqui, 'Les populations', p. 192; Ministère de l'Agriculture, *Statistique agricole de la France. Résultats de l'enquête décennale de 1862* (Strasbourg, 1868), p. xcix.

38. M. Block, *Statistique de la France*, 2nd edn (2 vols., Paris, 1875), vol. 2, p. 28.

39. Blanqui, 'Les populations', p. 192.

40. Ministère de l'Agriculture, *Statistique agricole, 1862*, p. xcviii.

41. Ibid., p. cv.

42. G. Duverger, 'Création d'une école régionale d'agriculture aux environs de Limoges', *L'Agriculteur du Centre: Bulletin de la Société d'Agriculture, Sciences et Arts de la Haute-Vienne*, vol. 1 (1848), p. 69.

43. Ministère de l'Agriculture, *Statistique agricole, 1862*, p. cxiii; M. Masclet, 'Etat présent de l'agriculture et de l'économie rurale dans le canton de Bourbonne-les Bains, Haute-Marne', *AAF*, vol. 2 (1829), p. 135.

44. F. Villeroy, 'Situation de la propriété foncière dans les provinces rhénanes', *JAP*, vol. 5 (1841-2), p. 166.

45. E. Labrousse, 'The evolution of peasant society in France', in E.M. Acomb and M.L. Brown (eds.), *French Society and Culture since the Old Regime* (New York, 1966), p. 59.

Symbols and Processes

No fewer than 7 million extra mouths were being fed in France in the mid-1830s by comparison with the end of the ancien régime and this was achieved predominantly from domestic sources of food. This feat was made possible by a combination of processes including land clearance, division of commonlands and large estates, increased production of established cereals and more widespread cultivation of relatively new crops such as maize and, more especially, potatoes. Each of these processes commenced before 1789 and the resulting agricultural achievements have been recognised as 'revolutionary' in their magnitude by members of what one may call the 'symbolic' school of historical interpretation who select symbols of the potential for agricultural change but often fail to demonstrate that productivity was raised. The work of Octave Festy provides a clear argument in favour of an early agricultural revolution so that, in his own words, 'one may take 1750 as the date when the new agriculture started to demonstrate its existence'.[1] It was at this time that a number of processes converged and began to generate an atmosphere in which change might be possible in limited sectors of French farming. A new enthusiasm for agriculture developed among people of more comfortable social status than the peasantry, that is the nobility, rich and poor, and possibly some successful bourgeoisie who were intent on becoming members of that class'.[2] English experiments in agricultural practice were publicised across the Channel in books and pamphlets to which a small number of French landowners had access. The physiocrats proposed their principles, believing that the wealth of any country existed in proportion to the fertility of its land. They were opposed to the excess and luxury of urban living and the evils of manufacturing and defended the virtues of agricultural life instead.

In the final decades of the ancien régime provincial agricultural societies were set up, first at Rennes (1757) and then in other parts of France, in order to spread the agricultural gospel of new crops (especially green fodders), livestock, fertilisers, agricultural implements and land clearance. The *Société Royale d'Agriculture* that was established in Rouen in 1760 undoubtedly had an important influence on introducing the new agriculture in the Pays de Caux before the Revolution but it was more the exception than the rule.[3] Rieffel could report that the agronomists of Rennes produced an important series of memoirs but failed to exert their influence on the province at large.[4] A few agricultural schools were established at the same time to foster similar objectives but these were scarcely feasible so long as communal systems of agricultural organisation survived over much of France. Official support was given to many innovations but there is precious little evidence that such information was diffused from the cultivated few to the cultivating masses or, even if it were, that the bulk of farmers had the means to act upon it. Certainly Arthur Young could report that in general, the tillage of the kingdom is most miserably performed; and many of the provinces are, in this respect, so backward that to English eyes they appear to be pitiably conducted'.[5] Bourde concluded that the impact of English ideas on French farming was less a fact of economic history than a fact in the history of ideas . . . This may explain why the success of the movement was limited in practice, while very great in theory'.[6] Even Festy, who argued that the new agriculture was already an 'organised force' by 1789, admitted that subsequent change was slow and spatially uneven, affecting different regions and forms of cropping in various ways; and hence 'decisive progress took place only in the second half of the nineteenth century in response to a range of new and favourable circumstances'.[7]

A more acceptable, dynamic approach to the study of French food production in the eighteenth and nineteenth centuries has been developed in recent years. This may be typified by Weber's masterly work on the modernisation of rural France.[8] Scholars have pursued different lines of enquiry but their work shares the common requirements of identifying processes whereby agricultural productivity might be raised and demanding proof that it actually occurred. Such investigations tend to be process-orientated and emphasise, for example, the diffusion of information about new agricultural techniques and the construction of adequate means of transport for surpluses to be carried from rural producers to urban consumers. Highway improvements and canal building in the early-nineteenth century played an important role in enhancing transport systems at the regional scale but not until after 1840 was a national system of communications gradually forged by the construction of railways, to link region to region, and of feeder roads, along which farm produce might be moved to main roads and stations. Rail and road in combination formed the most active agent encouraging agricultural progress. Since the first nationwide set of agricultural statistics dates from no earlier than the 1830s it is not surprising that dynamic interpretations of the so-called agricultural revolution' in France envisage it as being virtually synonymous with the railway age.

Augé-Laribé and Goubert accepted the localised importance of earlier changes but insisted that 'the real and unmistakable revolution came with the railways, which broke down barriers, shrank distances and unified the nation'.[9] Kindleberger adopted similar logic to reach an identical conclusion; while Henri Sée recognised an acceleration in agricultural productivity after 1840 but firmly placed 'decisive changes', such as the triumph of artificial meadows and the extension of fodder crops that made fallowing unnecessary, into the second half of the century.[10] In a somewhat different vein, Laurent has written of a 'continuing revolution' from the eighteenth century through to the twentieth century which affected different crops and practices in different regions with varying intensity and at different times; but, even so, it was not until after 1850 that 'each region became free to assume its natural vocation and be devoted to the most viable forms of cultivation'.[11] The hungry forties were experienced in France as well as in other parts of Western Europe and demonstrated not only the fragility of existing agricultural systems but also the inability of transport networks to move food supplies to where they were needed. Michel Morineau is the most implacable critic of the notion of an early 'agricultural revolution' and has painstakingly assembled literary and statistical evidence in order to justify banishing use of the term before the mid-nineteenth century.[12]

In the heat of the debate scholars focused their attention on the second halves of the eighteenth and the nineteenth centuries. As a result, the period between 1800 and 1850 failed to receive the attention that perhaps it deserved. However, Léonce de Lavergne recognised the peaceful years between 1815 and 1847 as 'the great epoch of national agriculture' and two recent pieces of work have argued a case for accelerating agricultural productivity prior to the 1840s or 1850s.[13] Toutain's ingenious calculation of national indices of agricultural productivity from various estimates and more sound statistical sources identifies the period 1815-24 to 1855-64 as displaying the fastest increase in agricultural output during the whole of the nineteenth century.[14] More recently, Newell has elaborated Toutain's interpretation by using official estimates, but estimates none the less, of the annual production of leading crops in each département after 1815.[15] As Juillard has remarked, the 'agricultural revolution of the eighteenth century is no more than a myth' but there are plenty of fragments of statistical and literary evidence to show that French agriculture was far from unchanging in the first half of the nineteenth century.[16]

One way of increasing food production without raising agricultural productivity or necessitating any kind of agricultural revolution, was to extend the surface under cultivation. Land clearance had been included in the list of official encouragements for agriculture in the final decades of the ancien régime but just how much heath, moor and marsh was brought into permanent cultivation between 1761, when fiscal encouragement began, and the 1830s remains unknown. Neckar estimated that 950,000 *arpents* were reclaimed between 1766 and 1780 but it is likely that much of that total comprised land on which clearance was intended but not achieved, at least within that time span. [17] According to Castang a mere 300-350,000 ha were brought into cultivation between 1761 and 1789, when *c.* 25,000,000 ha were under the plough and *c.* 10,000,000 ha left uncultivated. [18] Other estimates suggest an arable total of *c.* 21,000,000 ha with wastelands covering an area nearly half that size. [19] Revolutionary legislation in 1792 and 1793 for the division and sale of common grazing land was greeted with disfavour by the humbler peasantry who depended on the commons for raising a few head of livestock. Undoubtedly some land was reclaimed but further division was halted in 1795-7 and the legislation was rescinded in 1803 although confirming sales that had taken place already. There is no way of telling with any accuracy how much wasteland was cleared in subsequent decades but contemporary documents from all parts of France mention a progressive erosion of moors and heaths as pressure of population on land resources increased.

Clearance schemes operated at a wide range of scales, sometimes involving individual plots but also covering wide areas of rough grazing land which were often intersected by regular lattices of tracks and field boundaries. Squatters carved smallholdings from the waste and some villages and hamlets replaced their communally regulated pastures by owner-occupied plots. Cultivation was extended into difficult environments where soils were poor and clearance and ploughing of sloping terrain served to accelerate erosion. For example, fresh clearances were made in the high Auvergne to grow rye but yields were pathetically low and much soil was simply washed away. [20] Comparable problems were produced in other uplands where not only moorland but also timber was cleared. Unfortunately just how much land was cleared and how the rate of *défrichement* fluctuated during the Empire, the Restoration and the July Monarchy remain unknown. However, evidence on population growth and land use in the 1830s suggests the kinds of pays in which fresh cultivation most probably took place on an important scale.

Increased Output 1815-50

The Revolution brought the feudal, seigneurial system to an end and enunciated the freedom and rights of the individual. Royal, ecclesiastical and noble estates were expropriated and in many cases were divided up and sold. Farm leases were renegotiated and extended and tithes and taxes on agricultural produce and tolls on all internal trade were abolished. The *Code Rural* (1791) spelled out that 'the whole territory of France is free, as are the people who inhabit it'.[21] Each individual, in theory, had the right to cultivate his land and dispose of his crops as he thought fit, without reference to neighbours or to ancient communal rotations, and he was free to enclose his land if he wished. However, in practice, he might encounter difficulties since communal regulations affecting farming activity were certainly not swept away overnight. Not only did areas of undivided commonland survive into the nineteenth century but so too did the centrepiece of the old collective system namely the rights of stubble grazing (*vaine pâture*) and inter-commoning (*parcours*), as well as gleaning, raking and vine gleaning. Attempts to resolve some of these issues recommenced in the 1830s and areas of commonland were sold off in Anjou and other parts of France but action was not general at this time.[22] In 1836 the Government encouraged the leasing out of commonland but even in progressive départements, such as Seine-Inférieure, the problem was not tackled until after 1840. Similarly in 1836 and 1838 the Chamber of Deputies prepared projects to suppress vaine pâture and parcours. Widespread opposition to parcours was recorded in many parts of France but most départements urged that the question of vaine pâture be treated with great caution.[23] In areas where communal rotations were still operative, cultivators were far from being the completely free agents that the Code Rural would suggest.

However innovations most certainly did occur, with relatively new crops such as maize, potatoes and fodders being produced much more widely. By virtue of its environmental requirements maize was grown predominantly in the south-west and after occupying a mere 20,000 ha in 1790 potatoes came to be produced in virtually all parts of France to cover c. 300,000 ha in the first decade of the new century as popular resistance was eroded by fear of starvation.[24] For example, the tuber was eagerly accepted following food shortages in 1812 and the surface devoted to it increased from 560,000 ha in 1815 to 920,000 ha in the late 1830s.[25] It formed an important new source of foodstuff for man and beast but was subject to marked fluctuations in harvest. After 25,597,940 hl in the poor year of 1815 output increased to 71,982,800 hl in 1835 and 102,000,000 hl in 1840. The highest rate of increase between 1815 and 1835 was in the North-West (up 612 per cent), which had produced only 722,000 hl in 1815 but 5,145,000 hl 20 years later. The East region increased its output from 3,206,000 hl to 17,799,000 hl (up 455 per cent) to overtake the North-East as leading producer. Rates of increase were slow in the North-East where the potato had been long accepted, and in the South-West, where maize performed a role very similar to that of the potato in other parts of France.

Unlike the potato and maize, fodder crops such as clover and sainfoin (artificial meadows) performed an indirect role in enhancing food supplies. Loosening of traditional rotations and spread of the process of enclosure both before and after 1789 allowed green fodders to be grown in some but not all parts of France. Provision of more fodder enabled a greater number of livestock to be raised, more manure to be returned to the soil and hence successive cereal yields to be increased. For this reason cultivation of artificial meadows incorporated fundamental changes in agricultural production and opened the way for substantial progress in grain production and in the quantity and quality of livestock that might be raised. None the less, the persistence of communal-grazing regulations presented a serious obstacle to the success of green fodders in many pays.

In addition to these new crops, the production and the productivity of many established crops increased substantially. As the leading bread-grain and an important element in domestic trade, wheat received particular attention in the annual estimates that were recorded after 1815. The surface devoted to it rose from 4,591,670 ha in 1815, to 5,338,000 ha in 1835 and 5,951,380 ha at mid-century. The wheat harvest of 1850 (87,986,780 hl) was more than double that for the bad year of 1815 (39,460,970 hl) and the average national yield rose from 8.59 hl/ha to 14.78 hl/ha. Harvests were exceptionally good in some years such as 1832, when 80,089,000 hl were harvested and national average prices fell from 21.85 F/hl (1832) to 16.62 F in 1833 and 15.25 F in 1834. But in harsh years the home product became scarce, prices rose and wheat had to be imported (Figure 3.1). The harvests of 1812, 1816, 1817 and 1845-6 provided classic examples of such a disaster. For example, output fell from 82,445,840 hl in 1844 to 71,963,280 hl in 1845 and only 60,696,960 hl in 1846. Prices rocketed from 19.75 F/hl in 1845 to 24.05 F in 1846 and 29.01 F in 1847. The remaining years of the 1840s were ironically productive and prices plummeted. Because of spatial variations in supply and demand and in the efficiency of transport before railways were built, there were very marked regional differences in commodity prices even in relatively 'good years'. The most marked contrast in wheat prices was between the high levels of the south-east, which never produced enough grain to meet demand, and the low prices of the north-east, where surpluses were produced as a matter of course.

For most years between 1815 and 1850 France both imported and exported wheat (including maslin with which it was incorporated in the statistics) although for 1822-6 and 1834-5 the

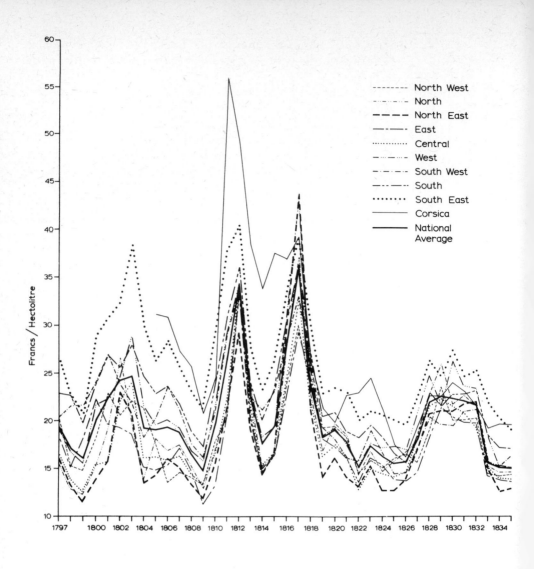

Figure 3.1: Average Wheat Prices, 1797-1835 (by Region)

volume of imports was very small. During five periods of varying lengths net exports occurred, ranging from 27-77,000 hl/pa in the mid-1820s to 1,500,000 hl in 1849. But such figures are small by comparison with the crisis years when imports greatly exceeded exports, for example, there were net imports of 4,168,000 hl in 1832, 4,781,000 hl in 1846 and 8,785,000 hl in 1847. However, the wheat trade cannot be treated in isolation from that of flour. For all but five years between 1816 and 1850 France exported more flour than she imported. In ten years no flour was imported but a net import of 785,000 hl in the bad year of 1847 contrasted with net exports of 1,030,000 hl in 1849 and 35,000 hl in 1850. When wheat and flour are considered together there was an equal division between importing and exporting years. For 1822-7, 1834-8, 1841-2 and 1848-50 France had wheat and/or flour to spare. In the remaining years she had to import either or both commodities, with combined net imports rising as high as 2,078,000 hl in 1817, 4,216,000 hl in 1832, 4,701,000 hl in 1846 and reaching a massive 9,570,000 hl in 1847. The following three years saw enormous exports of both wheat and flour. It is clear that the mid-1830s, immediately prior to the agricultural enquiry, were times of relative abundance when wheat yields were quite high, prices fairly low and sizeable amounts of wheat and flour were exported.

Wheat was the noble grain that figured in bourgeois diets and was eaten almost exclusively in some parts of the countryside but in many pays other grains were consumed. Taking France as a whole, an estimated 81.8 per cent of the wheat consumed in 1835 was used for human food, with a further 17.7 per cent employed for seed and a mere 0.5 per cent for livestock feed and 'industrial purposes' (Table 3.1). Other cereals were used in strikingly different ways but little was recorded about their production and trade at this stage.

National trends of cereal yields and total production were un-

Table 3.1: Estimated Consumption of Cereals, 1835 (%)

	Human Food	Livestock Feed	Seed	'Industrial Purposes'
Wheat	81.8	0.1	17.7	0.4
Maslin	82.2	0.9	16.7	0.2
Rye	79.7	1.6	18.2	0.4
Barley	50.6	20.3	16.1	13.0
Buckwheat	74.1	15.4	10.4	0.0
Maize	75.3	21.0	3.7	0.0
Oats	4.6	78.9	16.3	0.3
Pulses	71.3	16.5	11.8	0.4
All cereals	58.9	23.2	16.3	1.6

questionably upwards during the Restoration and the July Monarchy but there were important interruptions. Total output of cereals rose from 132,094,470 hl in 1815 to 204,165,190 hl in 1835 but each of the shares of that greatly enlarged total accounted for by maslin, barley, buckwheat, maize and oats declined. Surfaces devoted to wheat, barley, maize and oats increased in absolute terms, but those under maslin and probably also rye and buckwheat declined. The proportion of cereals used for human food fell slightly between 1815 and 1835; percentages kept as seed and used for industrial purposes increased marginally, but the proportion used for animal feed rose from 19 to 23 per cent. This trend was fairly consistent throughout the 20 years and reflected a 50 per cent increase in the number of horses and cattle which had been estimated by Chaptal as 2,000,000 and 6,681,950 respectively in 1812 but had risen to 3,000,000 and 9,936,500 by the late 1830s. [26] By contrast, the number of sheep declined from 35,000,000 to 32,151,430 over the same period, as areas of rough grazing contracted and fallows diminished. Alternative sources of animal feed, such as artificial meadows, were placed at a premium.

Little is known of regional trends in cereal production at this time but it is possible to compare estimates for 1815 with those for 1835. Output of wheat more than doubled in the west and Corsica and increased by 90 per cent in the north-east and the centre. Production of rye more than doubled in the centre, the east and Corsica, but output fell by eight per cent in the north-east. When all cereals are summated the highest percentage increase in output took place in Corsica (up 203 per cent) and the west (up 101 per cent). Lowest rates of growth were in the already advanced regions of the north (up 43 per cent) and the north-east (up 27 per cent). However, in terms of volume of cereals grown the north remained the undisputed leader at both dates, producing 35,902,000 hl in 1815 and 51,421,000 hl in 1835. The north-east occupied second position in 1815 (20,891,000 hl) and third place in 1835 (26,642,000 hl), having been overtaken by the north-west which produced 18,456,000 hl in 1815 and 28,360,000 hl in 1835. The three northernmost regions accounted for over half of total cereal output at both dates and displayed massive superiority over the remaining seven regions, which generated 43 per cent of the national harvest in 1815 and 48 per cent in 1835.

The Spectre of Shortage

In spite of substantial increases in food production during the Restoration and the July Monarchy, French agriculture remained unquestionably backward when compared with conditions

across the Channel. Lullin de Châteauvieux, the Swiss agronomist, travelled widely in both countries and argued that French
agricultural production had risen by one-third between 1789 and
the mid 1830s.[27] This was due to piecemeal improvement rather
than by generalised adoption of new systems of cultivation. He
claimed that French cereal yields were low, fallows widespread
and cultivation techniques rudimentary. Green fodders needed
to be grown much more widely if the quantity and quality of
French livestock were to be raised. Agricultural conditions only
in the extreme north of France could bear comparison with
English farming. French inferiority was due in part to the harsh
necessity of growing enough grains to ensure survival. This
characteristic had been inherited from the ancien régime and
was accentuated by the abolition of tithes at the Revolution and
the need to ensure sufficient domestic grain supplies during the
Revolutionary and Imperial Wars. In addition, links between
producers and consumers were much weaker in France than in
England and most French farmers were deprived of the immense
lever for improvement that catering for urban consumers
provided.

The spectre of food shortage haunted not only consumers but
also administrations during the Restoration and the July
Monarchy. As Kaplan explained for the ancien régime, 'the
subsistence problem dominated life . . . in a merciless and
unremitting way. No issue was more urgent, more pervasively
felt, and more difficult to resolve than the matter of grain
provisioning'.[28] Identical problems survived well into the new
century. Shortages provoked panic and civil disturbance as in
the north-east in 1816 and 1817 when harvests failed and
hardship was widespread.[29] Later food shortages produced
popular discontent in northern, western and central provinces in
1828-9, in all parts of the country except the south-east in 1832,
in the centre and west in 1839-40, and in many parts of France
and especially the west in 1846-7.

Attempts by the State to increase food production, control
prices and facilitate distribution took many forms. As Chapter 7
will show, importation and exportation of grains was regulated
in response to fluctuations in domestic harvests and commodity
prices. But even if supplies might be purchased from abroad they
would be of limited use to feed hungry mouths unless an efficient
system of internal transport operated between points of import
or areas of surplus and pays of shortage. The existing high road
system was certainly improved and the canal network was
extended in the early nineteenth century, however, legislation
for building and maintaining rural roads dated only from 1836,
and the railway network still had to be constructed at that date.
As a result, traditional methods of agricultural production and a

condition of virtual self-sufficiency continued to characterise much of France in the 1830s.

Notes

1. O. Festy, *L'agriculture pendant la révolution française* (Paris, 1947), p. 38.
2. G.E. Fussell, *The Classical Tradition in West European Farming* (Newton Abbot, 1972), p. 156.
3. A. Frémont, *L'Elévage en Normandie* (2 vols., Caen, 1967).
4. J. Rieffel, 'Institutions agricoles de l'Ouest', *AOF*, vol. 1 (1840), pp. 145-64.
5. A. Young, *Travels during the Years 1787, 1788 and 1789* (London, 1792), p. 128.
6. A.J. Bourde, *The Influence of England on the French Agronomes, 1750-1789* (London, 1953), p. 218.
7. Festy, *Agriculture*, p. 47, p. 440.
8. E. Weber, *Peasants into Frenchmen: the modernization of rural France, 1870-1914* (London, 1977).
9. M. Augé-Laribé, *La Révolution agricole* (Paris, 1955); P. Goubert, *The Ancien Régime: French Society, 1600-1750* (London, 1973), p. 59.
10. C.P. Kindleberger, *Economic Growth in France and Britain, 1851-1900* (Cambridge, Mass., 1964); H. Sée, *Histoire économique de la France: les temps modernes 1789-1914* (Paris, 1951), p. 137.
11. R. Laurent, 'Le secteur agricole' in F. Braudel and E. Labrousse (eds.), *Histoire économique et sociale de la France* (Paris, 1976), vol. 3 (2), p. 619, p. 694.
12. M. Morineau, 'Y-a-t-il eu une révolution agricole en France au XVIIIe siècle?' *RH*, vol. 486 (1968), pp. 299-326; M. Morineau, *Les Faux-semblants d'un démarrage économique en France au XVIIIe siècle* (Paris, 1970).
13. G. Léonce de Lavergne, *Economie rurale de la France depuis 1789*, 2nd edn. (Paris, 1861), p. 47.
14. J.C.Toutain, 'Le produit de l'agriculture française de 1700 à 1958', *Cahiers de l'ISEA*, vol. 115 (1961), série AF1, pp. 1-221.
15. W.H. Newell, 'The agricultural revolution in nineteenth century France', *JEH*, vol. 33 (1973), pp. 697-731.
16. E. Juillard (ed.), *Histoire de la France rurale: apogée et crise de la société, 1789-1914* (Paris, 1976), vol. 3, p. 9.
17. M. Necker, *De l'Administration des finances de la France* (Paris, 1784), vol. 3.
18. C. Castang, 'La Politique de mise en culture des terres à la fin d l'ancien régime', Unpublished thesis for Doctorat en Droit, Paris, 1967.
19. F. Quesnay, *Tableau économique* (Paris, 1758); H. de Turbilly, *Mémoire sur les défrichements* (Paris, 1760).
20. M. Lange, 'Rapport sur un ouvrage de M. Yvart ayant pour titre "Excursion agronomique en Auvergne"', *Mémoires de la Société d'Agriculture et de Commerce de Caen*, vol. 3 (1830), p. 71.
21. Augé-Laribé, *La Révolution*, p. 103.
22. Sée, *Histoire*, pp. 133-34.
23. M.L. Mounier, *De l'Agriculture en France d'après les documents officiels* (2 vols., Paris 1846), vol. 1, pp 209-14.
24. Toutain, 'Le produit', p. 94.
25. Ministère des Travaux Publics, de l'Agriculture et du Commerce, *Archives Statistiques* (Paris, 1837). (Département figures are summarised into nine mainland regions plus Corsica in this source); M. Block, *Statistique de la France* (2 vols., Paris, 1860), vol. 2.
26. J.A. Chaptal, *De l'Industrie française* (Paris, 1819).
27. F. Lullin de Châteauvieux, *Voyages agronomiques en France* (Paris, 1843).
28. S.L. Kaplan, *Bread, Politics and Political Economy in the Reign of Louis XV* (2 vols., The Hague, 1976), vol. 1, p. xvi.
29. J.D. Post, *The Last Great Subsistence Crisis in the Western World* (Baltimore, 1977).

The Diversity of France

The economic unity that had been proclaimed at the Revolution remained more illusory than real after the July Monarchy. Feudal dues and internal tolls had been swept away by the Revolutionary fiscal regime but systems of communication were highly imperfect and thereby prevented the practical implications of legal unity being experienced throughout the realm. Rather than being a truly unified country nineteenth-century France remained a collection of provinces and pays displaying strikingly different sets of environmental resources and food supplies (Figure 1.1a). Tradition and custom continued to permeate agricultural activities and, indeed, every aspect of life. Feudalism had been abolished in name but something very similar survived in many areas well into the nineteenth century. The new national systems of weights and measures were not accepted readily by countryfolk and provincial languages and patois remained the normal means of expression across wide stretches of rural France.

To quote Léonce de Lavergne, 'France is a summary of Europe and almost of the world'.[1] In addition, the agricultural inspectors and many contributors to contemporary farming journals confirm that the same kind of diversity that exists between regions may also be found within many individual départements. Rich and poor environments were frequently juxtaposed, with 'veritable oases being destined to satisfy the deficiencies of less fertile areas'.[2] The wooded and verdant interior of Charente-Inférieure contrasted strikingly with the fever-stricken coastal plains as though they were 'two different worlds'; the Limagnes 'shone like a diamond at the foot of the deserted Cantal'; Grésivaudan provisioned the Alps of Dauphiné, and Vaucluse supplemented the resources of the 'burning lands' of Provence.[3] Similarly the fertile valley of the Loire was surrounded by 'large heaths and desert plains' and, as Malte-Brun remarked, 'the rich banks of the Loire may be said to resemble one of those magnificent frames which deceive the ignorant and enhance in their opinion the value of a picture'.[4] The local populace identified scores of pays throughout France and in this respect the common sense of the peasants preceded science since it attributed a special name to each stretch of country with the same appearance and the same farming. As Lullin de Châteauvieux proved so admirably on his agricultural journeys, 'there is nothing homogeneous about French

47

agriculture. That is its most distinctive feature, of which one must never lose sight'. [5]

Faced with such variety, many administrators and agricultural writers in the nineteenth century recalled Arthur Young's attempts to produce a resumé of the environmental conditions that he witnessed on his travels during the 1780s. [6] The complexities of topography, geology, soil and natural vegetation which made up the face of the country were described and mapped as seven types of 'district'. In spite of their naïveté Arthur Young's map and descriptions were 'quoted with great reliance on their correctness' in many later works and formed the only comprehensive summary of the nation's land resources that was available during much of the July Monarchy. [7]

Such sparsity of knowledge occasioned an enquiry in the mid-1830s among prefects into the physical characteristics of their respective départements. [8] They were required to estimate the proportions of their territory belonging to ten different but unfortunately not mutually exclusive categories. Following Arthur Young's example, the enquiry referred to 'mountains' and 'heaths and moors' as well as eight more obviously pedological classes including not only 'rich soils' and 'clays' but also 'soils of different sorts'. It is not possible to produce a single composite map of soil types from this source since stretches of country were perceived by some prefects and their advisers as belonging to two or even three categories at the same time (e.g. stony soil, moor and mountain). Undoubtedly the quality of information at the disposal of individual prefects varied enormously so that their returns displayed bewildering cases of over- and under-statement. Administrators may have possessed a fair appreciation of local resources but they were unable to express that knowledge in suitable terms for a nationwide enquiry.

Lullin de Châteauvieux produced a classification of his own, which differentiated between soils that were 'fertile' (32 per cent of France), 'mediocre' (47.5 per cent) or 'sterile' (20.5 per cent). [9] The most fertile stretches were contained in territory conquered by Louis XIV along France's northern and north-eastern frontiers, namely in Nord and in the Rhine valley, with the Limagnes, parts of Isère, the environs of Avignon and Meaux, and certain valleys also being very fertile. By contrast, the poorest soils occupied a wide central area from Lozère to Finistère. However, the general lack of scientific knowledge was not to be rectified until the second half of the nineteenth century when treatises on agricultural geology were prepared and agricultural maps were drawn.

As well as discussing soils and geology, Arthur Young summarised France's climatic gradient by describing the supposed environmental limits for growing oranges, olives, maize and vines. The five climatic regions that he mapped were reproduced in various ways throughout the nineteenth century. The geographer Malte-Brun defended such an approach, arguing that 'the indications offered by different plants are less liable to error, and the climate of France may thus be better determined'. [10] This cartographic simplicity contrasts with Nicolet's *Carte de la géographie agricole de la France* which appeared in the *Atlas physique et de météorologie agricoles* in 1855 and on which the production limits on nine crops were shown, along with the major viticultural areas (Figure 4.1). [11] As well as typically southern crops that had figured on earlier maps, Nicolet showed the limits of northern crops such as hops, sugar beet and buckwheat. In addition, major areas of rye growing were sketched in the Massif Central, the Sologne, Champagne and the Landes. Seven years later an even more complex map appeared in Babinet's *Atlas universel de géographie* which attempted to depict twelve characteristic crops and two broader types of land use (heath plus marsh, and woodland). [12] Not until the final quarter of the nineteenth century were attempts made to map environmental and agricultural information in a quantitative

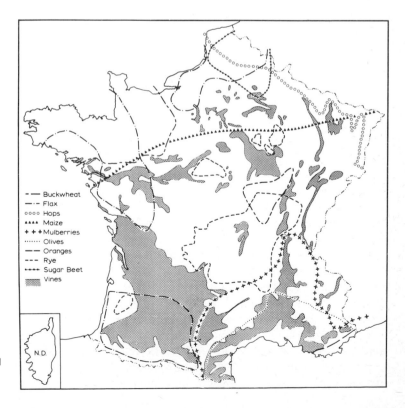

Figure 4.1: Agricultural Geography of France, after Nicolet

49

way. Gustave Heuzé's *Atlas, contenant une notice sur les régions agricoles* displayed information by département and represented a considerable advance on earlier agricultural maps.[13] But the data relate to a quarter century after the Statistique and portray a nation that had already entered the process of modernisation associated with the railway age.

At mid-century Adolphe Blanqui was able to assert that such a wide range of physical resources and agricultural production should, in theory, provide a guarantee of security for France. In several respects his claim appears reasonable. The greater part of the nation was composed of cultivators who produced enough food for their families to survive and to meet the needs of town dwellers and other non producers. In some years there was enough grain for export. In other respects the situation was much more sombre. Food shortages and acute distress, which incidentally had given rise to Blanqui's fact-finding mission, were registered in both town and country in the late 1840s. Failure of staple crops and the still primitive condition of internal transport in many provinces revealed the fragile agricultural base on which the national economy had been poised so precariously during the July Monarchy. Only a small minority of agricultural producers functioned as commercial suppliers to urban markets. Under Louis-Philippe over half of the French population lived in a 'closed economy' of local subsistence in which goods and services were exchanged and money was rarely used.[14] In the 1830s and 1840s deep-seated poverty was the normal state for the bulk of the peasantry and in bad years neither their security nor that of the nation as a whole could be guaranteed. For many countrymen life was much harsher than an inventory of environmental resources might suggest.

Lines of Communication

The task of improving internal communications to facilitate moving food supplies had been started during the final years of the ancien régime and continued through the early decades of the nineteenth century. However, the legacy of roads and navigable waterways inherited by the July Monarchy was highly imperfect. Rivers and ancient natural routeways retained much of their historic importance but imparted a very uneven and incomplete network of accessibility which, in turn, contributed to the degree of market orientation that might be expected from one pays to another (Figure 4.2). The Allier valley forms a fine example of a natural routeway cutting through otherwise mountainous terrain and channelling trade between Paris, Bourbonnais and parts of Languedoc. None the less, conditions were hazardous. Military surveyors complained that stones and building materials were in short supply in the Limagnes where

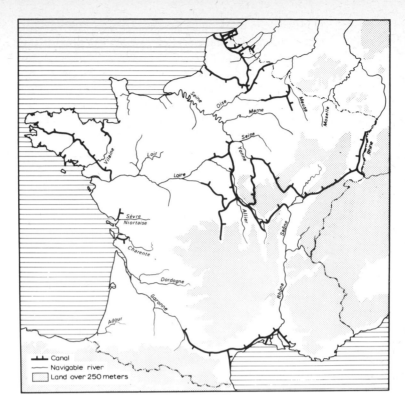

Figure 4.2: Canals and Navigable Waterways

Canal
Navigable river
Land over 250 meters

heavy rainfall interrupted movement.[15] Bridges were few and far between, fords were dangerous and roads degenerated into mere tracks as one moved upslope from the Allier. Things were even worse in Aveyron. Monteuil described the département as a cul-de-sac that could be entered by carriage only from the south-east or south-west; 'any other point of entry suggests coming into an enemy territory with which the government wishes to have no communication'.[16] Destruction of bridges by torrential flood water and threat of attack by brigands added to the difficulties facing travellers. By contrast, the legacies of physical geography and of a long history of administrative centralisation endowed the Paris Basin with a more complete system of major routeways than any other part of France.

In addition to recognising these basic facts it is necessary to examine the extent of road building and canal construction in the years immediately before the Statistique and attempt to gauge spatial variations in the degree of maintenance of the various lines of communication. New roads had, of course, been constructed in the late-eighteenth and early-nineteenth centuries and improvements had been made to existing highways. For

51

example, the intendants of Languedoc established a network of roads that were envied in other parts of France; opening of the Limoges/Bort/Clermont highway in 1786 reduced the isolation of the *montagne limousine*; while in Champagne new *routes royales* produced substantial improvements after 1750.[17] But the impact of such changes was simply linear and these regions remained relatively isolated in the 1830s. The same was true for Brittany where the Duc d'Aiguillon's highways had been built for moving troops but had little effect on the economy of the surrounding countryside.[18] Where routes royales were kept in good condition agricultural specialisation could take place, as around Isigny which successfully despatched butter along the Cherbourg/Paris highway during the eighteenth century.[19] None the less, contemporary observers were virtually unanimous in condemning the state of major and minor roads alike. Conditions were particularly defective in dissected terrain and in marshy areas that lacked local supplies of stone for building solid foundations.

Movement by water afforded a solution in some pays but the distribution of rivers and canals was uneven and conditions of navigation were poor in any case, with low water levels and the presence of shoals forming frequent impediments. Contemporary economists such as Michel Chevalier blamed France's industrial and agricultural backwardness on the lack of communications as well as inadequate systems of education and provision of credit.[20] He urged that new roads, canals and especially railways should be constructed with all haste since engineers were available on the staff of the *Ponts et Chaussées*, the *Mines* and the army. In his opinion all was ready in 1838 for a great development in the means of comunication. This was indeed to come after 1850 but France remained solidly in the pre-railway age in the 1830s.

The July Monarchy did form something of a turning point in the history of communications in relation to investment policy, classification of major roads, legislation regarding maintenance of rural roads, and finance devoted to strategic highways, canals and railways.[21] But by the time of the Statistique only a gentle start had been made and with respect to only some of these improvements. For example, only four railway lines (174 km) existed in 1837 and were of negligible significance for moving agricultural produce. Coal and iron were carried 20.5 km between Saint-Etienne and Andrézieux on the first line, completed in 1828 by Marc Seguin. No commodities were hauled upslope initially but by the mid-1830s sand, lime and timber were being carried from the river Loire to Saint-Etienne. That city was linked to Lyons in stages between 1830 and 1833 for the carriage of coal to the Rhône at Givors and to the Saône at Lyons/Perrache. With the exception of easing food supplies to Saint-Etienne the railways had no impact on domestic agricultural trade in the late 1830s.

The deplorable state of French country roads was recorded with monotonous regularity in travellers' journals and in publications of agricultural societies. The agronomist Mathieu de Dombasle insisted that minor roads needed to be funded and maintained more effectively if they were to play the vital role of feeding agricultural produce to main roads, canals and eventually railways. [22] In Nièvre the Nivernais Canal had been started and the Paris/Lyons highway and other roads had been repaired following severe food shortages in 1812 and 1816 but de Chambray believed that there was no chance of agriculture abandoning its old routines until minor roads were improved. [23] These were impracticable for several months each year and rendered essential tasks, like carting manure from cowshed to field, virtually impossible.

Legislation on rural roads was introduced in 1824 but had little effect. The state authorised communes to raise special resources for *chemins dits vicinaux* but did not compel them to do so. About 650,000 km were duly classified in the following twelve years and some roads were repaired but few new stretches were built. [24] A more forceful policy was required and this came with legislation of 21 May 1836 which introduced new rules concerning financial and physical responsibility for building and maintaining country roads. *Chemins vicinaux de grande communication* were differentiated from *routes départementales* and *chemins communaux* (or *chemins vicinaux de petite communication*). Communes and départements co-operated to construct this new category of roads which was placed under the direction of the départements. Able-bodied taxpayers were required to give three days' labour each year or pay an appropriate tax. It seemed that this might be difficult to implement but scarcely had the law been passed than the Minister of the Interior encouraged prefects and *conseils-généraux* to apply it. Local authorities responded with speed and a start had been made in designating chemins vicinaux in 82 départements before 1836 had elapsed. No less than 19,678,000 F were voted for the chemins vicinaux in a period of six months and Chevalier reported that enthusiasm was not waning in 1837, when substantial sums were allocated for new roads and bridges following the law of 14 May 1837. As a result of this kind of legislation the French countryside began to escape from its isolation, but the process was to be slow in the extreme and had only just commenced when the agricultural enquiry was undertaken.

By 1837, 3,699 km of canals were in existence and many major river basins had been inter-connected following construction of the Canal du Midi in the seventeenth century (Figure 4.2).[25] Seventeen départements contained over 100 km of canal apiece with the densely populated Nord (251 km) and the large département of Cher (249 km) at the head of the league. But absolute lengths can be misleading if they are not related to the areas that they serve. When that is done the small département of Seine (with only 22.8 km of canal) emerged as the best served with a location quotient of 6.8, being followed by Nord (6.3), Cher (4.9), Aude (4.2), and Haut-Rhin (4.1). Important work was in progress in the 1830s, such as the link between the Seine and the Rhine, but Chevalier could identify five main lacunae which needed to be remedied: Saône to Oise, Lower Seine to Lower Loire, Loire to Garonne, Bordeaux to Lyons (via the Dordogne and the Berry Canal), and from Garonne to Rhône which might only partly be solved by canals.[26]

Many existing canals were highly imperfect. For example, the Burgundy Canal has been opened in 1833 to provide a link between the Seine and Saône systems and allowed wheat surpluses from Côte-d'Or to be despatched either to Paris or to Lyons; however, Chevalier condemned it as being torrential for some periods of the year and virtually dry in others. Similarly, the Berry Canal, which was completed in 1830, was usable for only about 240 days each year because of ice or insufficient water. The Rhône-Rhine (Saint-Symphorien) Canal that was completed between 1802 and 1835 allowed northern grains to reach Lyons and southern wines to penetrate north-eastern France but it was criticised for frequent shortages of water and for numerous locks confronting boatmen. The banks of the Canal du Midi suffered frequent collapses and in Chevalier's eyes the watercourse only really merited the name 'canal' between Toulouse and the Mediterranean since nothing was 'more irregular than navigation along the Garonne to Toulouse'.[27]

Criticisms of this kind were raised for many stretches of the 8,964 km along 212 rivers that were deemed to be navigable or floatable. According to Chevalier 101 of these rivers flowed into the Atlantic, 42 to the Channel, 38 into the Mediterranean and 31 drained beyond the frontiers of north-eastern France. The most pressing problems involved the Garonne between Toulouse and Castets, the Seine between Troyes and Rouen, the whole length of the Rhône, which was described by Michelet as 'a raging bull', and the Loire upstream from its confluence with the Vienne where for half of the year the channel was graded into many small streams that flowed between banks and islands of sand.[28] An efficient lateral canal seemed to be the only solution and was in fact opened between Digoin and Briare in

1838. There were no fewer than 69 sandbanks along the Yonne 120 km downstream from Auxerre and boats could not descend unless a rush of water was released from the Nivernais Canal. Navigation on the Loire and the Isère was hindered by rapids, wrecks and varying depths of water, but the local population derived profit from hiring out horses to help the boatmen move their craft.

Even navigation on the Seine posed difficulties both upstream and downstream of Paris. Upstream navigation was halted for at least two months each year because of low water and movement downstream suffered from shifting channels and the presence of ancient and poorly maintained bridges, such as those at Vernon and Pont-de-l'Arche, which were encumbered with mills and offered serious hazards to boatmen. The first proposal for a railway to serve Rouen noted that the voyage by sailing ship between America and Le Havre took only three weeks but the journey to Paris through the meanders of the lower Seine required at least a month. Sailing the open sea offered few problems but the Seine was fraught with danger and between 1830 and 1852 no fewer than 500 vessels were lost between Tancarville and Caudebec. [29] 'Navigation on the Seine remained what it had been at the time of the Norman invasions'. [30] However, Chevalier alleged that nothing could be easier than to make the Seine navigable for deep-draught vessels throughout the year. [31] He believed that the river could readily become a model of navigability provided adequate finance was made available. A similarly devastating comparison was drawn in the 1820s between the speedy trans-Atlantic vessels plying between Bordeaux and Guadeloupe and the slow-moving coal boats which took several months to negotiate the waterways of northern France leading from the mines of Mons to the basin of La Villette in Paris. [32]

These and other qualitative deficiencies should be borne in mind when examining Figure 4.3 which relates the length of navigable rivers plus canals in each département to its total surface. The small and quite exceptional département of Seine had the best measure of navigable waterway (location quotient of 9.2). Other important foci included the confluence département of Seine-et-Marne (2.6), the canalised départements of Nord (3.7) and Ardennes (2.5), Bas-Rhin (2.7) and Haut-Rhin (2.2), the départements of the middle Loire and the Loire-Saône interfluve (Cher 2.2, Nièvre 2.1, Saône-et-Loire 2.1), and Maine-et-Loire (2.1). Five zones of under-provision emerge between these navigable axes: Champagne-Lorraine (with Vosges having neither navigable rivers nor canals); south-eastern France including Corsica; the Pyrenean départements; a vast stretch of central and central-western France (with Cantal, Corrèze,

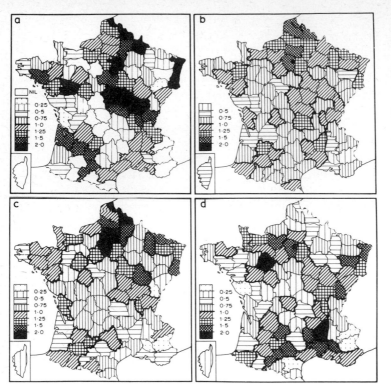

Figure 4.3:
Lines of
Communication:
Location Quotients
(a) All Inland
Waterways
(b) All *Routes Royales*
(c) *Routes Royales*
Maintained
(d) All *Routes
Départementales*

Creuse, Lozère and Haute-Vienne having no navigable waterways at all); and Lower Normandy plus Eure-et-Loir.

This kind of evidence must be interpreted with care. Firstly,
waterways directly served only a very small part of France hence
they must not be considered in isolation but must be examined in
relation to other forms of communication. Secondly, an index
that treats only inland navigation regardless of coastal movement is inevitably misleading. Movement by sea was easier than
movement by land along many stretches of coast. When the appropriate adjustment is made for coastal areas the fact remains
that fourteen land-locked départements were without any form
of navigable channel. Thirdly, the simple presence of canals or
navigable rivers is certainly not a surrogate for the quality of
those waterways or the intensity with which they were used.
The removal of internal tariffs at the Revolution also removed a
major element in the raison d'être for gathering statistics on inland navigation. As a result it is not possible to obtain a nationwide view of the volume of traffic passing along waterways.

Uncertainties concerning quality and intensity of use also sur-
round the question of communication on land. The issue is complicated further by changes in the administrative classification of

(e) *Routes Departementales*, Percentage Completed and Under Repair
(f) *Routes Royales* and *Routes Départementales* Maintained and Under Repair
(g) All *Chemins Vicinaux*
(h) *Routes Royales* and *Départementales* and *Chemins Vicinaux*

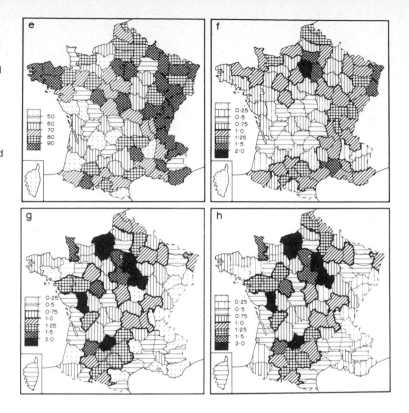

roads which occurred during the early nineteenth century. Road building did not fire the imagination in the same way as digging canals or laying railways but Chevalier insisted that it was the most urgent improvement since not a single inhabitant could fail to benefit.[33] The *Nouvelle carte routière de la France* revised by Simencourt in 1828 has been used as the basis for Figure 4.4 which displays the *grandes routes* as they would have been early in the July Monarchy. As might be expected, Paris stood at the focus of the national network and there was a clear contrast in the density of grandes routes between northern France, from Brittany to Alsace, and the remainder of the country. Some deficiencies were compensated by routes départementales or simple *traverses* which were shown on Simencourt's more complicated original map. For example, a score of poorly maintained traverses served the Causse du Quercy and in theory provided links with the Paris road which was lined by retaining walls to offer protection from rock falls and subsidence along steeply sloping stretches.[34] Even traverses were lacking in other pays such as the Landes, where the *route des petites Landes* had been paved by the Imperial administration at the time of the Spanish War but conditions remained appalling elsewhere.[35]

Simencourt's map sheds no light on the intensity of use to which main roads were put, their quality of construction and maintenance, or the various administrations that were responsible for them. Fortunately the introductory volumes of the Statistique contain a considerable amount of information for each département and enable road conditions to be examined in more detail. At the head of the hierarchy of road types came 34,512 km of nationally-funded routes royales in 1837, the greater part of which was declared to be well maintained. Only 3,134 km were paved but 21,583 km had a surface of stones. A further 5,863 km were classified as being in need of repairs (747 km to be paved and 5,106 km to be surfaced with stones) and the final 3,932 km that were necessary to complete the network still had to be built. Only 175 km of that final figure were to be paved, with the remaining 3,767 km to be surfaced with stones. The network of routes royales (also known as *routes impériales*) had been extended and improved in the middle years of the eighteenth century but also required much attention after 1800 because of the passage of troops or general lack of maintenance during the Revolutionary years.

As well as distinguishing between highways that were to be

the responsibility of the nation or the appropriate département, the Napoleonic decree of 1811 provided a classification of départements for purposes of road building. This emphasised the needs of areas traversed by main roads from Paris to Amsterdam, Mainz, Turin and towards Spain. Vendée was the only exception to that rule, with road construction being deemed a necessity to assist internal pacification and building the chef-lieu of Napoléon-Vendée (also known as Bourbon-Vendée and La Roche-sur-Yon). Cavoleau claimed that when all was completed the Vendée would be one of the most 'open' départements of France.[36] By contrast, land-locked départements such as Ariège, Gers, Corrèze and Creuse were treated like poor relations. In spite of progress along certain routes the Imperial period ended with a road network that was almost in the same state as it had been in the final years of the Revolutionary crisis. The energies of road engineers had been directed to the Alpine passes and to the Mediterranean coast, to afford links to Italy and to the Pyrenees en route for Spain. In all other parts of France the task of maintaining bridges, repairing degraded carriageways and building new roads still had to be started in the Restoration and the July Monarchy.

In 1828 Baron Jacqui reported to the Chamber of Peers that 14,288 km of the 33,536 km network of routes royales were severely damaged and required 67 million F to be put right.[37] A further 490 km remained to be finished and only 18,758 km were in relatively good condition. A start was made on constructing suspension bridges but the total achievement of the Restoration period was modest. By contrast, a much more dynamic transport policy was initiated during the July Monarchy, which was a time of external peace, economic revival and financial well-being, following the liquidation of war debts by the preceding regime. Construction of the canal system was actively pursued, the first railways were built and the road network was extended. Progress was slow in the first years of the reign of Louis-Philippe but the attempt by the Duchess of Berry in 1832 to arouse the western départements in support of her son, the Duke of Bordeaux, changed the speed of events. A network of strategic roads in eight western départements, to be built at the expense of the state, was approved in 1833. In all 38 routes totalling 1,462 km were built. Their military significance was virtually nil but they served an important purpose in opening up the west for agricultural change. In addition, legislation in 1837 started a programme of investment in routes royales which rose to 380,726,000 F in the years up to 1845. As a result, the routes royales increased from 33,536 km in 1824 to 34,512 km in 1837, 35,000 km in 1841 and 36,000 by 1855. The proportion deemed to be in good repair and fully usable rose from 43 per cent in

1824 to 72 per cent in 1837 and 98 per cent in 1855. These were things for the future, for in 1837 there remained a striking contrast in the distribution of routes royales, including sections under repair and still to be built. Départements in the Paris Basin were over-endowed but very low location quotients were recorded in the south and west. When sections under repair or still to be completed are excluded from the calculation the picture takes on a greater degree of reality, with Seine (3.3), Seine-et-Oise (2.2), Nord (2.1) and Oise (2.0) at the head of the league.

The second type of road comprised routes départementales which were established at the initiative of prefects and general councils and at the expense of individual départements. They varied greatly in quality and, for example, in Gard were little better than chemins vicinaux.[38] Most lacked drains or a stable surface and were particularly defective in the dissected terrain of the northern, *cévenol* part of the département. Early in 1837 the network of routes départementales extended to 36,578 km with 22,228 km (60.8 per cent) in good repair, 5,214 km requiring repair and 9,136 awaiting completion. Figure 4.3d conveys a rather diffuse display with Seine (7.4), Sarthe (2.5), Ardèche (2.1), Eure, Tarn, Vaucluse and Bas-Rhin (each 1.9) having the highest location quotients. The whole pattern needs to be interpreted in the light of three observations. First, location quotients were relatively low in the strongly agricultural parts of southern, central and western France, which is what one might expect from the topographic map. Second, location quotients for routes départementales were also quite low in parts of the Paris Basin and northern France, however these areas were well endowed with routes royales and/or navigable waterways. Third, the designation of another category of road (the chemins vicinaux), following the law of 21 May 1836, provided routes départementales in disguise. These chemins vicinaux were established by engineers employed directly by the département authorities and not by the staff of the Ponts-et-Chaussées. By early 1837 chemins vicinaux had been started in some départements that had been laggards with respect to other types of highway.

A corrective to the impression given in Figure 4.3d is afforded by Figure 4.3e on which the percentage of routes départementales constructed and under construction in early 1837 is plotted. Over 90 per cent of the network was complete or in progress around the capital, along the eastern frontier and in Brittany, but proportions were much lower in the south and south-west, with Corrèze (31 per cent), Corse (19 per cent) and Deux-Sèvres (7 per cent) coming at the bottom of the league. When routes royales and routes départementales that were either in good condition or were under repair are plotted a clear contrast in effective highway provision emerges between the Paris Basin, northern, eastern and Mediterranean France on the one hand, and the less well served départements of western, central and Alpine France on the other.

By early 1837 a network of no less than 771,458 km of chemins vicinaux was reported. Unfortunately no indication was given of the proportion actually completed but one suspects that it was low. Great lengths of chemins vicinaux were being planned (if not yet actually constructed) in already well served départements of the Paris Basin (e.g. Seine-Inférieure 3.3; Yonne 2.8; Eure 2.7) and also in western and central départements such as Deux-Sèvres (3.6) and Lot (2.8), which were under-provided with other forms of highway. Figure 4.3h summarises all types of road considered so far (routes royales and départementales and chemins vicinaux) irrespective of their state of completion or repair. It therefore reflects routes that were projected early in 1837, as well as roads that were in existence.

The existence of differing densities of means of communication is no substitute for a measure of the intensity of interaction. Unfortunately the absence of data on internal trade means that no more may be done than identify the potential for transporting foodstuffs rather than demonstrate movement. Figure 4.5 provides a basic summary of such potential by combining scores of 1.0 or more derived from inland waterways, routes royales plus routes départementales (in good condition and under repair), and chemins vicinaux. Départements that failed to reach the provision of even one form of communication that might be expected from the amount of land they contained were found in southern Champagne, Maine, the Alps, the central Pyrenees and land-locked territory from Corrèze to Loir-et-Cher. Six coastal départements also failed to score but navigation reduced their effective isolation. Many other départements in central and western France scored with respect to only one of the three means of communication and when that was chemins vicinaux (which were still largely projected) one may assume that the degree of isolation in 1837 was almost as great as for those départements with no score at all.

Only fourteen départements were well provided with each kind of communication and thereby achieved the maximum score. They were arranged from Saint-Etienne to Dunkirk in a distorted wedge which broadened in its upper part to incorporate most départements of the Paris Basin. The urbanised départements of Rhône and Seine and the agricultural département of Oise failed to score with respect to chemins vicinaux and broke the continuity of the wedge. Haute-Garonne and Indre-et-Loire plus Sarthe formed small outliers in the western half of

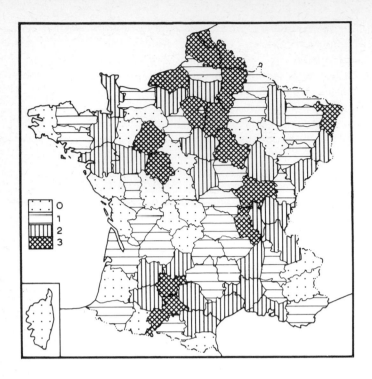

Figure 4.5:
Combined Index of
Roads and Waterways

France. The fourteen départements contained many of the richest soils and largest cities in the country. These were the areas where one might reasonably expect agricultural systems to be adjusted, in part at least, to meeting the needs of urban consumers.

Overtures

France was a predominantly agricultural nation during the July Monarchy. As Reclus remarked 'the peasant was master' and according to Schnitzler it was 'from the entrails of the earth that France drew her greatest riches'. [39] For most countryfolk farming was the way of life which enabled body and soul to be kept together but a minority of holdings were managed to meet needs other than those of the farming family and its neighbours. Demand and supply, rather than subsistence and survival, were the watchwords of their activity. As well as promoting improvements in transportation, Louis-Philippe's administrators established a series of organisations to encourage the diffusion of new agricultural ideas and techniques in order to enhance food production. In 1831 the administration of agriculture was rearranged and general councils were set up, not only for agriculture but also for commerce and manufacturing. [40] Working sessions began in 1832 and in that year a start was made to establish agricultural committees in cantons throughout France. These, in turn, organised agricultural shows (comices) at which prizes were awarded for notable agricultural improvements in each locality. Such grass-roots committees and larger agricultural societies received government support and financial backing and had the chance of spreading the 'sacred fire' of improvement into the farmhouses and fields which were 'the only possible laboratories for agriculture'. [41] According to Henri Sée they were 'quite active' and undoubtedly benefited from the return to the land, the so-called émigration intérieure by 'legitimist' land-owners following Louis-Philippe's coming to power.

In addition, government assistance for the new Ministry of Commerce and Agriculture (founded in 1836) was forthcoming for establishing model farms and providing official instruction in agriculture. Already a start had been made during the Restoration when the Royal Agronomic Institute at Grignon near Versailles had been set up to give practical and theoretical instruction in farming on 470 ha of royal estate that had been granted by Charles X. By 1840 Lullin de Châteauvieux could record no fewer than 157 agricultural societies, 664 committees and comices, 22 model farms (some with schools attached) and 15 special farm schools or agricultural penitentiaries. [42] Undoubtedly the impact of such organisations was limited in their early years but they represented a start in the vital task of diffusing information by practical example and by the spoken and written word.

The facts of physical geography, the cultural legacies of past generations and the efforts of contemporary administrators combined to attribute a spatial logic to the mosaic of agricultural production in the 1830s. If they were to expand major cities needed to draw on the resources of surrounding farmlands and perhaps also those of more distant pays. Commercial farming therefore developed in immediate proximity to demand centres and along axes of efficient communication by land or water. The Ile-de-France, Flanders, Dauphiné and the pays of the Rhône, Saône and Rhine valleys exhibited the necessary characteristics and may be expected to have developed the most intensive forms of production. Provincial cities normally drew the bulk of their foodstuffs from within a 15 km radius. [43] In the case of the capital more distant food sources of supply could be tapped along the upper courses of the rivers of the Paris Basin so that 'Paris was the sun in 120 km of atmosphere' away from the city. [44] Pays along the river Saône contained an important range of resources that might be transported to help satisfy the needs of Lyons and the cities of the lower Rhône. Similar, albeit smaller, examples of commercial food-production might be anticipated in parts of the basins of the Garonne, Loire and Moselle and in some coastal areas. By contrast, more isolated pays lacked such facilities for internal trade and may be expected to contain a predominance of semi-subsistent producers who exhibited limited participation in the money economy, geared their activities to meeting local needs, and had limited incentive for raising their agricultural productivity.

Notes

1. G. Léonce de Lavergne, *Economie rurale de la France depuis 1789* 2nd edn. (Paris, 1861), p. 60.

2. A. Blanqui, 'Les populations rurales de la France en 1850', *APAPER*, vol. 24 (1851), p. 194.

3. Ibid., p. 208, p. 194.

4. M. Malte-Brun, *Universal Geography* (Boston, 1833), vol. 8, p. 310.

5. F. Lullin de Châteauvieux, *Voyages agronomiques en France* (Paris, 1843), p. 519.

6. A. Young, *Travels during the Years 1787, 1788 and 1789* (London, 1792).

7. Anon., *Review of the Agricultural Statistics of France* (London, 1848), p. 4.

8. Ministère des Travaux Publics, *Statistique de la France: Territoire et Population* (Paris, 1837), pp. 90-95.

9. Lullin de Châteauvieux, *Voyages*, p. 112.

10. Malte-Brun, *Universal Geography*, p. 185.

11. 'Atlas physique et de météorologie agricoles' Paris, 1855. BNCC Ge DD 1265.

12. M. Babinet, 'Atlas universel de géographie', Paris, 1862. BNCC Ge DD 1853.

13. G. Heuzé, 'Atlas contenant une notice sur les régions agricoles' (1875). BNCC Ge DD 1125.

14. J. Vidalenc, *Le peuple des campagnes: la société française de 1815 à 1848* (Paris, 1970), p. 363.

15. M. Michaux, 'Le département du Puy-de-Dôme en 1841-2 d'après les mémoires et reconnaissances militaires', *ACSS* (1963), pp. 829-41.

16. A.A. Monteil, *Description du département de l'Aveyron* (Paris, 1803), p. 220.

17. L. Dutil, *L'Etat économique du Languedoc à la fin de l'ancien régime* (Paris, 1911), p. 896; A. Demangeon, 'La montagne dans le Limousin', *AG*, vol. 20 (1911), pp. 316-37; G. Arbellot, 'La grande mutation des routes de France au milieu du XVIIIe siècle', *AESC*, vol. 28 (1973), pp. 765-91.

18. J. Chombart de Lauwe, *Bretagne et Pays de la Garonne: évolution comparée depuis un siècle* (Paris, 1946), p. 83.

19. A. Frémont, *L'Elevage en Normandie* (2 vols., Caen, 1967), vol. 1, p. 62.

20. M. Chevalier, *Des Intérêts matériels en France: travaux publics, routes, canaux, chemins de fer* (Paris, 1838), p. 11.

21. J.C. Toutain, 'Les transports en France de 1830 à 1965', *Cahiers de l'ISEA, Economies et Sociétés*, vol. 1 (8) (1967), pp. 1-306.

22. C.J.A. Mathieu de Dombasle, *Des Chemins vicinaux en France* (Paris, 1833).

23. M. de Chambray, 'De l'agriculture et de l'industrie dans la province de Nivernais', *AASAF*, vol. 13 (1834), pp. 4-22; G. Thuillier, 'Les transformations agricoles en Nivernais de 1815 à 1840', *RHES*, vol. 34 (1956), pp. 426-56.

24. H. Cavaillès, *La Route française* (Paris, 1946).

25. Ministère des Travaux Publics, *Statistique de la France: Territoire et Population*, pp. 40-53.

26. Chevalier, *Intérêts matériels*, p. 59.

27. Chevalier, *Intérêts matériels*, p. 64.

28. Cited in F. Rivet, *La Navigation à vapeur sur la Saône et le Rhône, 1783-1863* (Paris, 1962), p. 26.

29. L. Girard, *La Politique des travaux publics du Second Empire* (Paris, 1952), p. 18.

30. J. Vidalenc, *Lé Département de l'Eure sous la monarchie constitutionnelle, 1814-48* (Paris, 1952), p. 539.

31. Chevalier, *Intérêts matériels*, p. 65.

32. M. Chevalier, *Cours d'économie politique, fait au Collège de France* (Paris, 1842), p. 325.

33. Chevalier, *Intérêts matériels*, p. 30.

34. W. Charra, 'Notes sur l'évolution des Causses de Quercy au cours du XIXe siècle', *RGPSO*, vol. 20 (1940), pp. 175-221.

35. H. Cavaillès, 'Le problème de la circulation dans les Landes de Gascogne', *AG*, vol. 42 (1933), pp. 561-82.

36. J.A. Cavoleau, *Statistique ou description générale du département de la Vendée* (Fontenay-le-Comte, 1844), p. 306.

37. Cavaillès, *La Route française*, p. 199.

38. H. Rivoire, *Statistique du département du Gard* (2 vols., Nîmes, 1842), vol. 1, p. 282.

39. E. Reclus, *Nouvelle géographie universelle: la France* (Paris, 1881), p. 881; J. Schnitzler, *Statistique générale*, vol. 3, p. 20.

40. H. Sée, *La Vie économique de la France sous la monarchie censitaire, 1815-48* (Paris, 1927), p. 37.

41. J. Rieffel, 'Institutions agricoles de l'Ouest', *AOF*, vol. 1 (1840), p. 149.

42. Lullin de Châteauvieux, *Voyages*, p. 77.

43. C. Clark and M. Haswell, *The Economics of Subsistence Agriculture* 4th edn. (London, 1970), p. 192.

44. M. D'Harcourt, 'Réflexions sur l'état agricole et commercial des provinces centrales de la France', *AASAF*, vol. 1 (1829), p. 259.

5

Components

At the dawn of the present century Vidal de la Blache wrote that 'the word that best characterises France is variety'.[1] His choice was unquestionably apposite but the historical geographer who reads the record of the French land and people on the eve of the railway age is tempted to insert the qualification 'bewildering'. Such a sense of bewilderment and even frustration arises from trying to assimilate a rich but fragmented testimony. Most contemporary evidence emphasised the unique ecological blend of spatial, physical and cultural conditions in individual localities and pays. This kind of approach undoubtedly encapsulates reality but also renders broad comparisons and the formulation of generalisations virtually impossible.

At the other extreme, some observers managed to reduce the complexity of rural life to a number of basic contrasts, for example, between feudal tradition, grande culture, tenancy, heavy horse-drawn ploughs and triennial rotations in the north, and the legacy of Roman law, petite culture, share-cropping, light ploughs drawn by oxen and biennial rotations in the south.[2] Plains could be contrasted with uplands and the diversity of France could be reduced to a distinction between bons and mauvais pays. This concern for dualities produces a greater degree of simplification than should satisfy the historical geographer. The avalanche of figures contained in the Statistique allows him to do more by displaying the 'average' situation in each arrondissement by means of quantitative maps which convey at a glance the range of conditions that relate to that level of generalisation.

Two dozen crops, or types of land use, were recorded in the agricultural enquiry of the Statistique together with information on various kinds of livestock. Their patterns of production, productivity, consumption and price will be examined in subsequent chapters, but first it is appropriate to ask what was the look of the land that managed to meet most food requirements of a rapidly growing population during the July Monarchy and supplied other essential commodities such as firewood. To bring the discussion within manageable bounds individual crops and land-use features have been grouped into six components: woodland, natural grass, arable, permanent crops, garden crops and 'other' land uses.

The first two were self-selecting but the others involved manipulation of original data. The arable component is derived by summating the surfaces under all types of grain crop, artificial meadows, potatoes and fallows. 'Permanent crops' were taken to include vines, chestnuts, olives, mulberries, madder and walnuts; while 'garden crops' comprised gardens, hemp, colza, flax, beet, tobacco and hops. This classification is, of course, open to debate, since crops such as potatoes or beet were grown only in gardens in some parts of France but entered field-crop rotations elsewhere. Unfortunately the Statistique does not draw that distinction but it does provide information on the value of production/ha for each crop and this, together with an index which compares value of production with the surface involved, has been used to help decide whether an item should be classified as an arable crop or a high-value crop (Table 5.1). Even so, ambiguities remain, with chestnuts undoubtedly being a 'permanent crop' but being worth much less than vines or olives. The final component has been described as 'other' forms of land use and represents the total of all remaining elements in each arrondissement. It refers almost exclusively to moors and heaths that were labelled together as *pâtis, landes, bruyères* in the département totals but were not mentioned for arrondissements. For this reason the 'other' category of land use had to be determined by a process of elimination. National dimensions of land use have been derived from the Statistique and are shown in Table 5.2, where they are set alongside roughly corresponding categories derived from the ancien cadastre but it must be remembered that the cadastre was drawn up for fiscal purposes and did not employ the same land-use classification as the Statistique. It did not describe conditions at a single point in time but spanned roughly four decades, nor was it complete at the date of the first agricultural enquiry.

To grow enough grain to ensure the survival of more than 33,500,000 people was the prime aim of French agriculture in the early years of the July Monarchy and half of the land surface was devoted to that end. As Madame Romieu remarked in 1865 'our fathers regarded cereal growing as the essential objective of farming to which everything else was subordinate'.[3] Moors and heaths covered no less than one-fifth of France, with only a slightly smaller proportion being under trees. There were some fine pastures but they were rare since all natural grasslands involved less than one-tenth of the country. One-twentieth of France was covered by permanent crops, with the remaining 1.7 per cent being under garden crops.

Each crop and form of land use owed its distribution to the interaction of distinctive aspects of the human and physical environments at particular points in space. For example, natural grasslands required year-round moisture in order to flourish. Such conditions were found in well-watered mountains, along perennial watercourses and in many coastal locations but not in

Table 5.1: Land-use Components

	Area (ha)	Index of Value/Surface	Value of Production (F/ha)
Wheat	5,586,786	2.19	197.40
Spelt	4,733	1.84	171.10
Maslin	910,932	1.75	158.25
Rye	2,577,253	1.27	114.95
Barley	1,188,189	1.28	115.85
Oats	3,000,634	1.11	100.65
Maize	631,731	1.26	113.65
Buckwheat	651,241	1.04	94.25
Pulses	296,925	1.94	175.15
Artificial Meadows	1,576,547	1.43	129.25
Potatoes	921,977	2.43	219.21
Fallows	6,763,281	0.15	13.05
Vines	1,972,332	2.69	212.30
Chestnuts	455,386	0.33	29.70
Olives	121,586	2.08	198.09
Mulberries	30,438	15.77	1,405.45
Madder	14,674	7.25	636.65
Walnuts	6,742		
Gardens	360,696	4.83	435.55
Hemp	176,148	5.44	489.85
Colza	173,506	3.27	294.65
Flax	98,241	6.52	585.35
Beet	57,663	5.58	502.55
Tobacco	7,955	7.93	689.30
Hops	826	12.50	1,191,25
Natural Grass	4,198,197	1.22	110.20
Crown Woodland	52,972	0.21	52.75
State Woodland	1,048,907	0.34	31.35
Communal and Private Woods	7,333,965	0.26	23.55
Sol forestier	368,705	0.00	0.00
Pâtis, etc.	10,191,076	0.09	8.95

the Midi, where pronounced summer drought necessitated irrigation for growing grass, nor on the *limon*-topped plains overlying permeable rocks in the Paris Basin which, after the harvest was gathered in, gave the appearance of a treeless steppe. By contrast, the extensive lowlands of northern France and Aquitaine, together with the smaller basins of the Midi, Alsace and the Limagnes were much more suited to various forms of grain growing. Differences in local environmental potential could be accommodated by variations in crop, strain and rotation but it was almost as if regions such as the Paris Basin were 'destined by nature for cereal cultivation', as Lullin de Châteauvieux suggested.[4]

Table 5.2: Basic Land-use Categories

	Statistique de la France (ha)	(%)	Ancien Cadastre (%)	Statistique Agricole, 1892 (%)
Arable	24,110,229	47.63	50.30[1]	48.83
Permanent Crops	2,601,158	5.13	5.43[2]	4.32
Garden Crops	875,035	1.73	1.23[3]	1.30
Natural Grass	4,198,187	8.29	10.14[4]	11.70
Woodland	8,804,550	17.40	17.35[5]	18.69
'Other' Land Uses	10,025,803	19.81	15.54	15.15
Total	50,614,972			

Notes: 1. *Terres labourables* 2. *Vergers, jardins* 3. *Vignes, oliviers, châtaigniers* 4. *Prés* 5. *Bois, forêts et domaines, oseraies.*

That was not the case since cultural factors had important roles to play. The general absence of natural grassland in Languedoc was not due solely to climatic factors but also to the widespread belief that fallows, moors and heaths could supply adequate fodder for the necessary livestock. Indeed, such reasoning operated across many other parts of France and hence livestock numbers were low, animals were poorly nourished, quantities of manure to fertilise the soil were small, and resultant cereal yields were low. In addition, high densities of human population and limited possibilities for transporting food meant that cereals were grown in hazardous environments where yields were low and harvests often failed. Large areas of the pays known as the *marais* of Vendée were devoted to cereals even though damp air and frequent rain made harvesting 'extremely precarious'.[5] The Pyrenees, Franche-Comté, the Massif Central and all upland areas supported ploughland at elevations and in fragile environments that would now be unthinkable.

Permanent crops such as the vine and the olive had quite specific environmental requirements which excluded them from cooler parts of northern and upland France. Virtually all regions grew garden crops to some extent but they required intensive inputs of labour and proximity either to demand centres or to efficient axes of communication if they were to be produced in large quantities. Peri-urban locations were therefore ideal. Moors, marshes and heaths encompassed a wide range of ecological conditions and were to be found on poor soils at all elevations and in both particularly damp and excessively dry locations. For example, the so-called 'isle of the Camargue' was inundated for much of the year but dried out hard in summer so that it became 'almost a desert. The soil was parched and cracked . . . A stranger coming across the land would think he

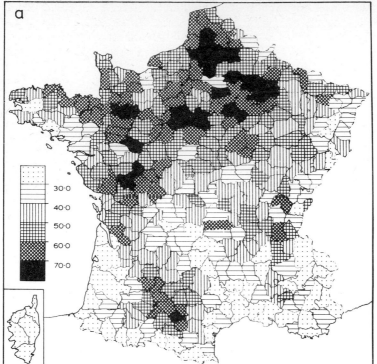

Figure 5.1:
(a) Ploughland and
(b) Natural Grassland

30·0
40·0
50·0
60·0
70·0

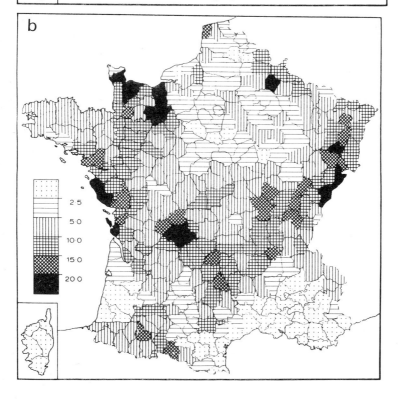

2·5
5·0
10·0
15·0
20·0

was in the midst of the steppes of Asia or America rather than in a country where science, industry and the arts have reached such a high degree of perfection'.[6] Finally, woodlands had survived where they had been protected by special laws and customs or where population densities were low.

The internal composition of the 24,110,229 ha of arable land varied greatly between arrondissements, with respect to the crops that were grown and the amount of land that was left fallow each year. A great mis-shapen quadrilateral, in which ploughland made up more than half of the total land use, stretched across northern France between Côtes-du-Nord, Charente-Inférieure, Haute-Saône and Nord (Figure 5.1a). Few northern arrondissements had less than half of their surface under plough, with 21 having over 70 per cent of their land in arable use, especially in Picardy, Champagne, Brie and Beauce.

Beyond these great northern granaries there were four other important arable areas. The largest of these comprised the basin of the Garonne which was described by Blanqui as 'one of the most fertile areas, not only in France, but in the whole of Europe'.[7] The plain of Toulouse was the granary of the Midi and despatched important supplies to Mediterranean France. No less than 81 per cent of the land around Villefranche was under the plough. A spur of arable land projected eastward from the Paris Basin into Lorraine but Moll could complain in 1837 that too much of the Moselle valley was under the plough. He believed that it was well suited to irrigated meadows but the farmers were not interested since 'wheat was virtually the only crop they could sell'.[8] A fragmented axis of ploughland extended along the Rhône and Saône valleys, with important concentrations being found near Lyons in the Bresse and Dauphiné which were separated by the Dombes marshes. Ploughlands occupied much land in the basins of Forez and the Limagnes and were so rich by comparison with the surrounding mountains that Blanqui called them a *pays de cocagne*.[9]

Arable farming was of great importance in the Midi and in some localities everything else was subdued to it. But fragmented terrain, an alternation of upland and lowland, and the presence of permanent crops reduced the arable surface to less than one-third in many southern arrondissements. Proportions fell below ten per cent in parts of the Pyrenees and the Alps and in two sections of Corsica; whilst only two per cent of Paris arrondissement was used for arable farming but there were still arable fields inside the walls as well as important market gardens. 'Middle France', arching round from Dordogne through Upper Poitou, Bourbonnais and Nivernais, supported extensive areas of marsh and moor on outwash materials carried down from the Massif Central and thus formed a zone of transition between arable openfields in the Paris Basin and very different assemblages of land use on rising ground to the south.

By contrast with wide expanses of ploughland stretching to the far horizon, very few arrondissements in the Paris Basin had more than five per cent of their surface under natural grassland and these were in well-watered vales and claylands such as Boulonnais, Bray, Bessin and Auge (Figure 5.1b). Elsewhere in the Paris Basin meadows were only found along watercourses since dry summers excluded the possibility of successful grassland farming. In some instances they were badly managed but they were usually held in high esteem because of their rarity value. Thus the meadows of the Lys valley were believed by Lefebvre to be the most expensive land in Nord.[10] Grassland gained significance in the land-use pattern with distance from the heart of the Paris Basin, to cover 10-15 per cent or even 20 per cent of a ring of arrondissements stretching from Cotentin, through the marchlands of Armorica, Vendée, Limousin, Cantal, Charolais, Franche-Comté, the Vosges and the Ardennes. André Siegfried evoked the contrast between the 'naked horizons of the ploughlands and the mysterious labyrinth of green meadows, veiled with trees and divided by innumerable hedges' as the traveller entered the marchlands of western France.[11] Environmental conditions in these pays were extremely varied, ranging from coastal marshes and damp vales in the west to high mountain pastures in central and eastern France, and from clays and volcanic soils to those derived from granitic parent-materials.

The amount of land devoted to natural grassland declined substantially beyond this pastoral ring as more arid environments in the Midi were approached, with every arrondissement of Provence, Languedoc and Corsica having less than 2.5 per cent of its surface under grass. The mountains of the central Pyrenees and the well-watered Gave de Tarbes contained the only sizeable areas of natural grassland in the south. Parts of Lower Normandy, with long-established links to the Paris market, had begun to be put down to grass in the late-seventeenth century and formed France's pastoral region par excellence during the July Monarchy, but only Lisieux and Pont-l'Evêque arrondissements had as much as one third of their surface under grass. No other area reached so high a proportion, although six more arrondissements had between 25 and 29 per cent of their land under meadows (Bayeux, Alençon and Argentan in Lower Normandy; Saint-Yrieux in Limousin; Pontarlier in Franche-Comté; Rethel in the Ardennes). By comparison with ploughland, which covered virtually half of France, the eight per cent under grass was small indeed.

Four types of woodland were identified in the Statistique;

Figure 5.2:
(a) Woodland and
(b) Other Land-uses

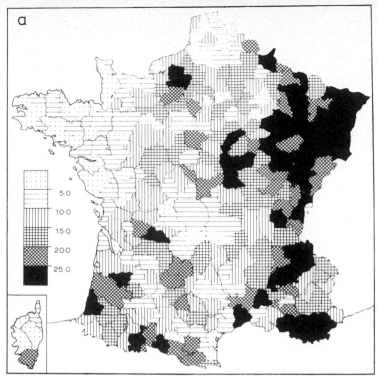

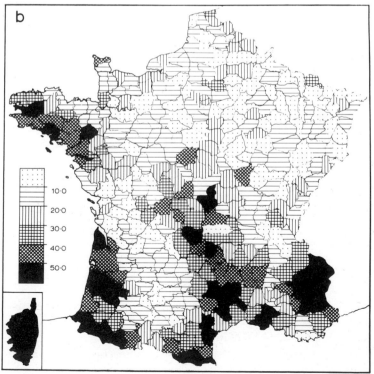

private and communal timber were treated together and covered 83.3 per cent of the total; state woodland involved 11.9 per cent; Crown woodland 0.6 per cent; and unwooded stretches of *sol forestier* within woodland perimeters 4.2 per cent. More than a quarter of the land surface of north-eastern France was tree-covered, with the sole exception of arrondissements in the Moselle valley (Figure 5.2a). High in the Vosges, no less than 51 per cent of the arrondissement of Saint-Dié was tree-covered. Further south, the Morvan, Dauphiné, Provence, the Pyrenees and the Landes were extensively wooded, with eighteen arrondissements having over 35 per cent of their land under trees, double the national average. State and Crown forests, which had been subject to special legislation, accounted for the survival of relatively well-wooded stretches in the Paris Basin. But just as striking as the presence of woodland in eastern France was its relative absence from Armorica and from smaller areas in Champagne pouilleuse and in the densely populated pays of Flanders and Artois.

Permanent crops required to stay in the ground for more than one year and sometimes for many. By definition, they were not included in rotations although seasonal crops might be grown between them. Vines accounted for about three-quarters of the total areas in this component, with the other permanent crops also being predominantly southern in location (Table 5.2). Land defined as *jardins* in the Statistique made up two-fifths of the garden crop category, with hemp and rapeseed together contributing a further two-fifths (Table 5.2). Beet covered 6.6 per cent and the very high-value crops of tobacco and hops only 1.0 per cent. Each of these crops had its own requirements for production and displayed a distinctive distribution. Not surprisingly, garden crops predominated in densely-populated areas with the three arrondissements of Seine devoting over ten per cent of their surface to garden crops (Paris 14.3 per cent, Saint-Denis 13.9 per cent, Sceaux 11.6 per cent). Even higher proportions were recorded in Nord (Lille 20.9 per cent, Douai 14.8 per cent) and Pas-de-Calais (Arras 16.8 per cent, Béthune 16.0 per cent).

'Other' land uses represent the amount of land remaining in each arrondissement when all other categories listed in the Statistique have been subtracted. They included a wide range of ecological types from high moors to low heaths, marshes and scrub and were located in four major blocks: central and southern Brittany; Corsica; the Landes, Pays de l'Adour and western Pyrenees; and a vast delta-shaped area from Prades to Montluçon and Castellane, including the greater part of the Massif Central, the Alps and the Camargue (Figure 5.2b). Isolated but very distinctive outliers were found in the Morvan, the Sologne and other *mauvais pays* in middle France. More than half of the land surface of thirty-five arrondissements was covered by 'other' land uses but they were of minimal significance in the arable landscapes of the Paris Basin, in the polycultural systems of Aquitaine, and in north-eastern France where ploughlands were separated by stretches of timber.

Contemporary opinions were diametrically opposed on the utility of the heaths, moors and marshes that still covered one-fifth of France during the July Monarchy. For the lesser peasantry they provided vital resources to supplement small holdings. Livestock could be grazed on the heather and fine grass that was produced in regions such as the Landes after rougher vegetation had been burned off. But Ponts-et-Chaussées engineers argued that the practice was dangerous and had led to disastrous forest fires in 1803, 1822 and other years.[12] In any case they maintained that there was rarely enough grass for the sheep during the winter and that livestock starved if low-lying pastures flooded because of heavy rainfall. In their opinion afforestation after drainage offered the only rational use of land in the Landes. Moors and heaths could also be used for periodic cultivation and food production or until the soil became exhausted. Brushwood could be gathered for firing, whilst herbs, berries and wild game afforded useful additions to meagre diets. Vegetation could be cut as litter for animals or dug in as a fertiliser in areas where animal manures were short. The latter was the case in the Pays de l'Adour where cutting the *touyas* in spring and autumn formed important activities in the agricultural calendar.[13] But in the eyes of the agricultural inspector these moors offered a 'sad spectacle' and led to the 'dishonour of the département of Hautes-Pyrénées'.[14] Jules Rieffel lamented that the equivalent of two whole Breton départements remained uncultivated because of lack of capital, initiative, fertilisers and settlers to colonise them.[15] In addition, it was often unclear who held legal title to the moors. Some belonged collectively to whole communes, *sections*, residents of particular hamlets, or indeed to individual landowners. Herein lay the major defence of the old order. The pages of *Agriculture de l'Ouest*, the transactions of the *Association Bretonne* and all agricultural journals offered advice on the choice between défrichement and afforestation but legislation to simplify land ownership and accelerate clearance was not passed until after the July Monarchy ended.

Provinces

By using a standard statistical technique to combine the information relating to the six basic categories it is possible to recognise a series of land-use provinces. The approach involves

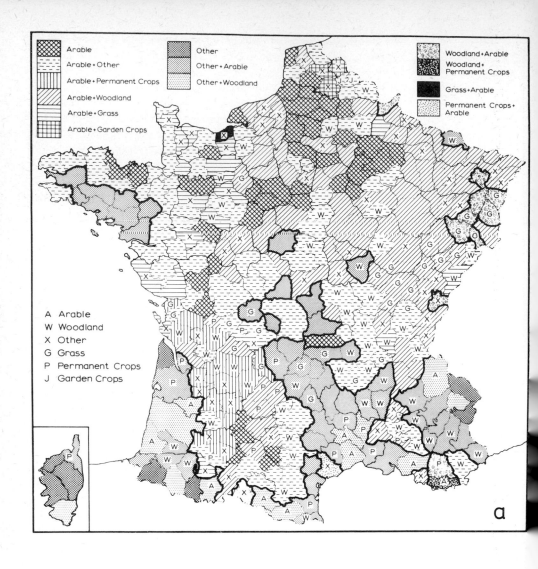

Figure 5.3: (a) Land-use Combinations and

b

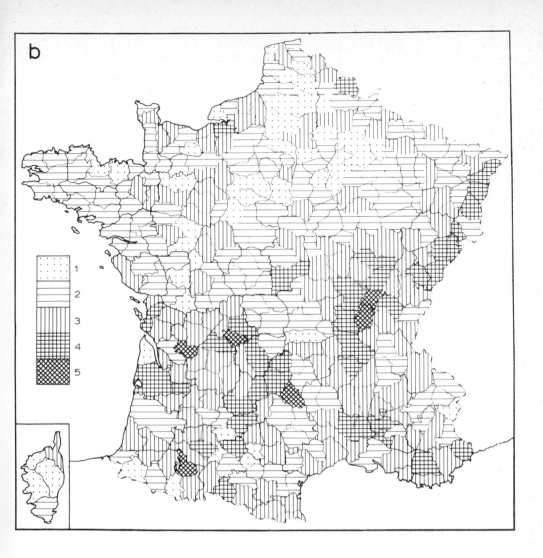

1
2
3
4
5

(b) Number of Components

77

the least-squares method of defining crop combinations that was used by J.C. Weaver and modified later by D. Thomas. [16] For each arrondissement the percentages of land occupied by each component are arranged in descending order to produce a distinctive curve. The shape of the curve for any arrondissement may be compared with theoretical curves for 1, 2, 3, 4, 5 or 6 components using the sum of the square of the differences between the percentage of the land devoted to each component and the corresponding point on the theoretical curve. The theoretical curve that matches most closely each arrondissement is deemed to be the best description of the land use of that area.

The resulting land-use combinations have been mapped in slightly simplified form (Figure 5.3a). It is obviously possible for as many as six components to merit inclusion by the least-squares technique but for purposes of cartographic clarity no more than three components have been depicted on the map. In fact, land-use descriptions involving more than three components were generated in only 46 arrondissements which were located in Alsace, Franche-Comté, the valleys of the Rhône and Saône, and in parts of the south-west (Figure 5.3b). They comprised areas where permanent and/or garden crops entered significantly into the land-use balance and, in other words, tended to reflect the distribution of the vine. By contrast, many parts of the Paris Basin and some sections of the Garonne valley displayed land-use combinations that were best described by a single component, namely 'arable'. A similar situation was recorded in central Corsica but there the dominant use of land was 'other uses'.

The results derived from generating land-use combinations for each arrondissement are shown on Figure 5.3a which records, as appropriate, the leading one, two or three land-use components. The one or two largest elements are shown by differing shadings, with the third element being depicted by literal symbols. A heavy line has been inserted to distinguish provinces that have been defined by their leading land-use component.

The 'other' (or landes) province involved no fewer than 77 arrondissements, distributed in the four main blocks described above, with outliers in middle France. By far the most widespread combination in this province was 'other and arable', which covered the whole of central and southern Brittany and much of the Massif Central and the south-east. However, 'other and woodland' characterised fourteen arrondissements mainly in the Landes, Pyrenees and Languedoc. Permanent crops appeared as the third element in the Bordelais and Languedoc, with grassland being the third element in sections of the pastoral mountains in and around the Cantal. Woodland was the leading component in only twelve arrondissements found in the Jura,

Ardennes, Pyrenees, Provence and the Vosges, with the latter zone set in a wide region characterised by 'ploughland and woodland'. Only in the Pays d'Auge (Pont-l'Evêque) did grass occupy leading position in the land-use combinations and permanent crops led only in Béziers arrondissement.

The remaining 272 arrondissements made up the 'arable' province, which spread over virtually the whole of the northern half of France, including not only the Paris Basin but also the marchlands of Armorica, northern Brittany, middle France and two prolongations southwards into the Basin of Aquitaine and along the Rhône valley. There were six sub-types with distinct regional locations within this vast area. 'Arable' alone formed the dominant land-use component in Picardy, Champagne, Beauce, northern Brittany, parts of Poitou, the Limagnes and sections of the plains surrounding Toulouse. 'Arable and woodland' provided the typical land-use combination of north-eastern France, with other stretches covering parts of the Ile-de-France, Upper Normandy, the middle Loire and Lot. An unbroken zone from the Charentes through Dordogne and the middle Garonne to Gers was characterised by 'arable and permanent crops', whilst 'arable and garden crops' involved only the three contiguous arrondissements of Lille, Béthune and Arras. 'Arable and grass' was rarely encountered apart from in Lower Normandy and a few sections of the Vendée, Pyrenees and Franche-Comté. By contrast with these combinations, 'arable and other' extended across wide areas of France and fringed what has been identified as the province of landes (or 'other' land uses). Thus the greater part of Armorica, the north-eastern, north-western and southwestern margins of the Massif Central, parts of Champagne and virtually the whole of middle France contained landscapes that involved a mixture of the two apparently diametrically opposed forms of land use, arable and waste. But, of course, the contrast was less sharp than it might at first seem since landes might be reclaimed for short periods of temporary cropping and extensive sections of arable might be fallowed for several years.

As befits a nation in which grain was the pilot sector of the economy rural landscapes of France on the eve of the railway age were characterised predominantly by arable land, with other major land uses being included to varying degree according to the local environment and the relative openness of individual areas enabling them to supply commodities to demand centres. In some, less commercial, contexts polyculture remained the order of the day and peasant farmers produced a wide range of foodstuffs to meet local needs. The information contained in this chapter should be seen only as the basic framework into which the more detailed elements of cropping and the rural economy need to be placed. Five themes, in particular, need investigation.

First, the extent of land occupied by individual crops and forms of land use needs to be identified with respect to the various components recognised in this chapter. Second, the production and productivity of these crops needs to be investigated, for example, with reference to seed: yield ratios for cereals. Third, the relationship between production and consumption of commodities needs to be plotted, with areas of surplus and shortage and variations in price being recognised for livestock products as well as for crops. Fourth, the varying composition and food value of regional diets may be examined to offer a useful index of the well-being of the peasantry; and finally a measure of the financial yield of cropping, animal husbandry and forestry needs to be devised in order to synthesise spatial variations with respect to each aspect of production.

Notes

1. P. Vidal de la Blache, *Tableau de la géographie de la France* (Paris, 1905), p. 40.

2. G. Léonce de Lavergne, *Economie rurale de la France depuis 1789*, 2nd edn. (Paris, 1861), p. 317.

3. Mme Romieu, *Des Paysans et de l'agriculture en France au XIXe siècle* (Paris, 1865), p. 171.

4. F. Lullin de Châteauvieux, *Voyages agronomiques en France* (Paris, 1843), p. 139.

5. J.A. Cavoleau, 'Description du département de la Vendée', *AAF*, vol. 3 (1818), p. 375.

6. AN F 10 209a. Desséchements. Bouches-du-Rhône. Rapport au Comité du Salut Public par la Commission d'Agriculture et des Arts (dated 4 prairial an II); D. Pigéaire, 'Excursion agronomique dans le Midi de la France', *JAP*, vol. 6 (1842-3), p. 550.

7. A. Blanqui, 'Les populations rurales de la France en 1850', *APAPER*, vol. 24, (1851), p. 202.

8. L. Moll, 'Voyage agricole en Lorraine', *JAP*, vol. 1 (1837-8), p. 260.

9. A. Blanqui, 'Extrait de la relation d'un voyage dans le Midi de la France', *ASLIA*, vol. 2 (1829), p. 3.

10. G. Lefebvre, *Les Paysans du Nord pendant la Révolution française* (Paris, 1924), p. 200.

11. A. Siegfried, *Tableau politique de la France de l'ouest sous la troisième république* (Paris, 1913), p. 4.

12. AN F 10 2341. Landes de Gascogne. 'Note sur l'assainissement et la mise en culture du département des Landes' by Lessore (date 20 November 1848).

13. M. Bottin, 'Statistique agricole, arrondissement de Pau', *Le Cultivateur*, vol. 8 (1833), pp. 121-23, 244-54; S. Lerat, *Les Pays de l'Adour* (Bordeaux, 1963).

14. Inspecteurs de l'Agriculture, *Agriculture française: Hautes-Pyrénées* (Paris, 1843), p. 87.

15. J. Rieffel, 'Assemblée générale du congrès à Nantes', *Bulletin Association Bretonne* (1843), p. 10.

16. J.C. Weaver, 'Crop combination regions in the Middle West', *Geographical Review*, vol. 44 (1954), pp. 175-200; D. Thomas, *Agriculture in Wales during the Napoleonic Wars* (Cardiff, 1963).

Rotations and Fallows

With such a wide range of environmental resources and no less than half of the national territory under the plough during the July Monarchy striking spatial differences are to be expected in the distribution and association of the twelve 'crops' that made up the arable realm. Each crop responded to ecological and economic factors to display its own discrete pattern of production. Physical parameters were obviously of great importance in creating these patterns but for four main reasons they may be discussed in only a cursory fashion. First, contemporary literature made little mention of the precise strains that were grown and hence one remains largely ignorant of the detail of tolerance limits. Second, the complexity of environmental resources defies adequate synthesis for some kind of index to be applied meaningfully to production figures for each arrondissement. Third, the cultural context of crop production was poorly recorded. Standardised information on man/land relationships in the 1830s simply does not exist and it is impossible to determine with any precision how many people worked the land, the types of tenure under which they operated or the size of holdings that they farmed. Hence the characteristics and intensity of farming systems may be indicated in only a partial way. Fourth, the degree of openness of particular pays to market forces remains largely a matter of conjecture, although simple location, information on land and water routes, and differences in commodity prices provide some clues.

By contrast with a sparsity of information on the agro-ecological environment, the crops themselves and the rotations according to which they were grown were recorded quite fully. An examination of rotations offers an organising framework whereby the diversity of land use may be approached. Crop innovations, such as potatoes, maize and green fodders, were documented relatively well as instigators of agricultural change. They reduced the amount of land left fallow each year, provided valuable sources of food for man and beast, and added to the complexity of arable practices. The present chapter will examine where innovative crops and long-established cereals were grown, how they were rotated and how they may be associated statistically in the form of crop combinations.

In some instances crop combinations and rotations were virtually identical. For example, a two-element 'wheat and fallow' combination or a three-element 'wheat and oats and fallow' combination summarise respectively the classic biennial and triennial rotations. But in other instances crop combinations only offer clues and do not give precise descriptions of rotations. A six-element combination might reflect a six-yearly rotation but it might just record that at least six crops were being produced as part of a modified biennial or triennial rotation. Crop combinations are simply statistical summaries of all the patterns of cropping within specified areas. In arrondissements dominated by single rotations they may be fair reflections of those rotations but such is not the case in areas where several systems of production were to be found.

The explanatory power of the biennial and triennial rotations that operated across wide areas of France fired the imagination of Marc Bloch and Roger Dion in their masterly syntheses of the processes that fashioned her agrarian systems and produced her rural landscapes.[1] In addition, it was recognised that poor soils in marginal environments were often used irregularly and phases of cropping were interspersed by varying lengths of long fallow. Yields would fall unless land could be fertilised adequately or rested in such a way. In any case land that was cropped continuously by the same cereal was typified by miserable outputs and outbreaks of plant disease. Equal alternation of phases of cropping and fallow (during which the land could be worked before seedtime) provided the simplest solution and such a biennial rotation was characteristic of the southern half of France and a few locations further north. The ecological features of the Midi, in particular its parched summer months, restricted the range of cereals that might be grown (Figure 6.1a). Drought and heat caused precocious maturation and hence spring grains were rarely sown since they would have had a precarious existence in a short season threatened by the dry summer heat. Consequently it was necessary to concentrate on winter corn, especially wheat and to a lesser extent rye, which could be sown in autumn. Preparing the soil took several months, with ploughing beginning in April, May or June following completion of harvest in the preceding July or August. The system took fourteen to sixteen months from the first ploughing to the final harvest and under such an agricultural calendar the land could be sown only once in every two years. The absence of pronounced summer drought in northern France permitted another form of rotation to be practised, with a spring cereal being inserted between winter corn and fallow. Preparation for spring cereals began at the beginning of the winter and never lasted for more than eight or nine months from the first ploughing to the harvest. Under this system land was fallowed once in three years and yielded a *petite céréale* (such as oats, barley or rye) that was used partly for livestock feed, in addition to the leading crop

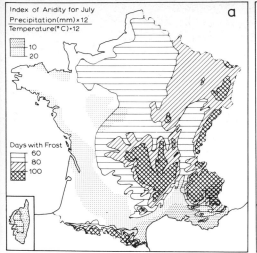

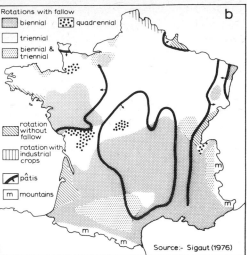

Reproduced courtesy of AESC, Paris-6^e.

Figure 6.1: (a) Aridity and Frost and (b) Rotations

that was largely for human consumption. As the prime bread-grain, wheat formed the leading element in both biennial and triennial rotations in many parts of France, and this fact helps account for its dominance in the national total of arable land (Table 5.1). However, other cereals (such as rye, maslin or buckwheat) assumed that function in areas with relatively poor soils or colder, moist climates.

The role of summer drought in the vegetative cycle thus gave the biennial rotation a predominantly southern distribution and the triennial rotation a mainly northern one. Faucher insisted that the latter required naturally fertile soils and would therefore be found on the limons of the Paris Basin and the loess-derived soils of Alsace.[2] The pioneer researchers did not proceed beyond providing word pictures of the typical environments in which basic biennial and triennial rotations might be expected. No attempt was made to map the distribution of these and other types of rotation until 1976 when François Sigaut published his sketch map which was derived from scrutiny of four hundred literary sources with a median publication date of 1820.[3] The main dividing line between biennial and triennial rotations ran from the Gironde around the northern edge of the Massif Central to Lake Geneva, with a major outlier of biennial rotations in the Lower Loire and smaller patches in parts of Normandy, the

83

Vosges and northern Alsace (Figure 6.1b). Biennial and triennial rotations co-existed in Aquitaine and parts of Brittany, Normandy, Burgundy and the Alpine fringes of Dauphiné, with an outlier of triennial rotations in the southern Massif Central. Quadrennial rotations were found in Lower Normandy and three stretches of middle France; whilst rotations without bare fallow were encountered in the Ardennes, Jura, Landes and southern and western Brittany. Intensive cultivation of industrial and fodder crops typified Flanders, where abundant supplies of fertiliser, good communications and the absence of collective rotations enabled many farmers to gear their production to what was in demand and to maximise their output. Finally, parts of Armorica, upland areas and much of middle France contained extensive moors and heaths where long fallowing was practised.

Two general observations need to be made in relation to both time and space. The sources used by Sigaut, while centering on 1820, actually span the years from 1750 to 1900. They do not relate to a precise period nor do they describe the ideal, supposedly pristine conditions outlined by the pioneer researchers. Source materials scrutinised for the present study span a shorter period but have a median date very close to that for the literature used by Sigaut. This fact may help account for some of the differences in detail that emerge. Sigaut favoured an impressionistic cartographic technique rather than one which recorded features precisely. His definition of moors and heaths (pâtis) was presumably inferred from the topographic map and from a rather wide definition of Armorica. The cartographic techniques he employed would appear to overstress distributions, notably in Flanders and the Grandes Landes, since many sources emphasised the specificity of ecological relationships between particular environments and rotations, but that kind of spatial subtlety could not be shown at the scale at which he was working. In any case, cartographic representation of literary references which often lack precise spatial definition is bound to be a hazardous business.

Examination of early-nineteenth-century sources suggests that Sigaut's map might be modified in detail and also raises a fundamental question regarding the value of simply defining rotations as biennial, triennial or whatever. For example, the agricultural inspector recognised two types of biennial rotation in Haute-Garonne: wheat/maize and wheat/fallow.[4] The first incorporated a new crop and reflected intensive farming but the second symbolised the traditional order. Even more confusing are references to fallows (*jachères*) which are qualified by such crops as artificial meadows, maize and potatoes which were grown on parts of the fallow land. Jachères, in the contemporary literature at least, did not always mean bare fallows. Sigaut's sketch might be modified in the following ways. Triennial rotations were described in Aude, Doubs and Maine-et-Loire, around Limoges, in the Ségalas of Lot, in Flanders, in parts of Puy-de-Dôme and possibly also in the Camargue. With the exception of Doubs, which was recognised as having a rotation without fallow, each of these areas was classified by Sigaut as having biennial rotations. Rotations involving long fallows were found in the Camargue and in the Champagne; whilst four-year rotations were recorded in Vendée and Haute-Saône and five-year rotations in Aveyron and mountainous areas of the southwest which did not support maize. Intensive cropping without fallows characterised parts of Lower Alsace and the Limagnes, as well as areas identified by Sigaut.

This information on rotations helps to establish a framework within which evidence on cropping from the Statistique may be set. Wheat may be expected to occupy a particularly dominant position in the composition of arable land in southern areas which had biennial rotations, whilst proportions of a third or less would suggest that it occupied the leading year of the triennial rotation that functioned mainly north of the Loire. Large expanses of oats may indicate the second year of the triennial rotation and extensive stretches of bare fallow are to be expected in marginal environments characterised by long fallowing. The half, third or quarter of the arable surface that was left fallow according to traditional rotations had been eroded by newly-introduced crops in many parts of France by the 1830s.

As well as a dozen types of arable crop, including those which had invaded the fallows, the Statistique recorded the surface that was left 'en jachères' each year. Hence it seems reasonable to assume that the official use of the term was not identical with that in many contemporary publications and did, in fact, refer to bare fallows. Morris Birkbeck underestimated the situation in 1814 when he reported that 'one fourth of the whole country is lying in a state entirely unproductive, a few weeds excepted', since two decades later no less than 28 per cent of French ploughlands was recorded as jachères and it is clear from contemporary observers that bare fallows had retreated considerably since the beginning of the century.[5] Greater freedom from communal systems of cultivation in many parts of France since the Revolution, division of some commonlands and large estates into smallholdings, and introduction of new fodders and subsistence crops were widely quoted to explain the decline of bare fallowing. Examples abound for many parts of northern France and to a lesser extent for the centre and south, but the pace of retreat varied significantly from area to area and even from commune to commune. In some areas farm leases continued to stipulate that a proportion of land be left fallow each year. Around Béthune (Pas-de-Calais) this was only one-ninth of the holding but in parts of Yonne, Doubs and Nièvre one-third of the farm had to be rested each year and 'this order of cultivation was invariable'.[6] In areas in Lorraine the custom of parcours continued to be practised and this, together with a lack of enclosures and the detailed requirements of leases, retarded the disappearance of bare fallows.[7]

Agricultural writers were fairly unanimous in their praise for the erosion of the fallow third in northern France and they eagerly anticipated the time when fallows would disappear completely. By contrast, commentators in southern France were more divided in their opinion. Small holdings, poor soils, summer drought and shortage of fertiliser were quoted as justification for the continuance of extensive bare fallowing in parts of Provence.[8] In the mountains of Var 'fallowing is not just a matter of routine, it is a veritable law in this area because of the operation of parcours. Progress is virtually forbidden'.[9] Others claimed that bare fallowing offered advantages for preparing the land and making use of two seasons' moisture for one year's crop; but some writers were unwilling to compromise and simply dismissed the defence of fallows on ecological grounds as 'banal and useless pretexts', justifying their opinion by reference to more intensive forms of Mediterranean agriculture in parts of Spain and Italy.[10]

In a score of arrondissements on the southern margins of the Massif Central, in Bourbonnais, Morvan, Marche and Upper Poitou 50 per cent or more of the ploughland was recorded as jachères (Figure 6.2a). This surely referred to the process of long fallowing rather than the unlikely survival of biennial rotations in unmodified form. Long fallowing, biennial rotations or a combination of the two may well explain extensive areas of bare fallows in upland regions (Massif Central, Upper Provence, Jura, Vosges) and areas of poor soil (Berry, marchlands of Armorica) where more than a third and often more than two-fifths of the arable surface was recorded as jachères. Only the arrondissements of Mauléon, Avranches, Saint-Lô and Strasbourg were reported to be without fallow land. Proportions of one-tenth or less occurred around Lille and Paris, in parts of Normandy, Léon, Morbihan, the Bresse, Angoumois, Périgord and the extreme south-west. Each of these areas had made early progress in adopting artificial meadows, maize or potatoes. By contrast, in Artois, Champagne, Perche and Upper Poitou between a quarter and a third of the ploughland remained as bare fallows and represented an only slightly eroded form of triennial rotation. Similar proportions in the Basin of Aquitaine and in the Midi méditerranéen represented biennial

Figure 6.2: (a) Fallows
and (b) Maize as
Percentage of Arable
Land-use

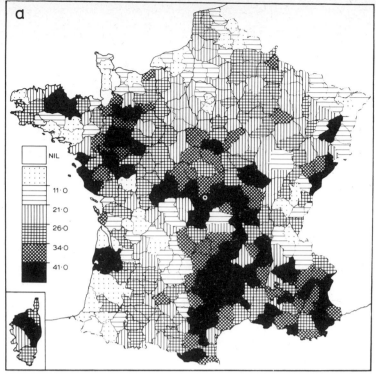

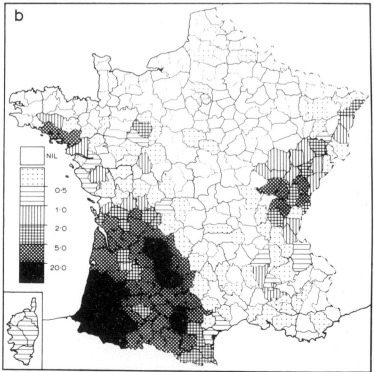

rotations in varying degrees of relaxation in response to crop innovations.

Crop Innovations

Taken together, maize, potatoes and fodder crops known as artificial meadows occupied one-eighth of French ploughland in the 1830s but their importance was greater than the area that they occupied would suggest. Each covered part of the traditional fallow period, required intensive cultivation, and provided important sources of food for humans, in the case of maize and potatoes, and for livestock, in the case of all three crops. However, they tolerated differing ranges of environmental conditions, were introduced and were accepted more widely at different periods and involved differing parts of France. Maize had very specific ecological requirements, being most suited to the extreme south-west of the country where it was already well established by 1800. Potatoes would grow successfully in virtually all parts of France except the Mediterranean south but they were accepted fully in only a few areas prior to the Revolution. Their adoption formed one of the key elements of agricultural change in the following half century, as did wider cultivation of the various fodder crops, each of which had distinct environmental preferences and histories of diffusion.

Maize covered 631,731 ha in the mid-1830s, a mere 2.62 per cent of the nation's ploughland, but occupied more than 40 per cent of the arable in six arrondissements in the extreme south-west, rising to 60 per cent around Bayonne (Figure 6.2b). In parts of the south-west the bare fallows of the old biennial rotation had been banished by maize cultivation but perhaps not as widely as Arthur Young had suggested, since the new form of biennial rotation was only found on very good alluvial soils or in relatively small and easily fertilised areas around settlements. [11] Much more widespread in Aquitaine was a triennial rotation of wheat/maize/fallow (with vegetables or green fodders on part of the fallow third) which replaced the old-established alternation of wheat and fallow. Secondary foci of maize cultivation were found in the *terrefort toulousain*, in Périgord, southern Morbihan and the Bresse. The crop was reported as far north as Béthune and Metz but it had no chance of ripening in such northerly environments.

Opinions differ on when maize was introduced to French territory. Various periods in the sixteenth century have been quoted for the Pays de l'Adour, Haute-Garonne and the plain of Alsace; whilst dates in the seventeenth century are cited for the environs of Toulouse and the valleys of the Adour, Gaves and Nestes in the Pyrenees. [12] Regardless of first appearances, widespread cultivation of maize in Aquitaine occurred in the eighteenth century with encouragement from agricultural societies, bishops and intendants. After 1750 maize was introduced from the south-west to the Rhône and Saône valleys, Alsace, and even the Loire valley and parts of the Paris Basin. As well as making use of former fallow land, maize could be intercalated with crops such as beans and beet and in some places replaced traditional cereals, such as wheat, rye and oats in the south-west and buckwheat in Alsace. It continued to make progress in the first half of the nineteenth century.

In order to ripen satisfactorily, the strains that were in use during the July Monarchy required a mean temperature in excess of 18°C between May and September.[13] This restricted the effective growing area to the south-west, the Rhône valley, parts of Alsace and the Mediterranean south, but the latter region was eliminated because of summer drought. In the areas where it succeeded, maize offered a number of advantages over other crops. It flourished on many types of soil and its success removed the need to import grains to feed the population of Béarn and other parts of the south-west. Yields were fabulously high, being two or three times those derived from other cereals, but maize was less nutritious than wheat, being deficient in protein.[14] None the less, as the eighteenth century progressed maize gradually became the subsistence crop par excellence for the peasantry and artisans of the south-west, thereby allowing wheat to become more of a cash crop that yielded income for payment of taxes. In order to achieve such rich returns the soil demanded heavy manuring and a great deal of attention, with three hoeings in each season being the norm. Large amounts of labour were required and hence maize growing was associated with areas of small farms and high densities of rural population. It was a classic 'cleaning' crop and yields were enhanced from the wheat and other crops that succeeded it in rotation. Its growing season was only between 70 and 80 days and so even if a harsh winter had ruined autumn-sown wheat there was still time to plant maize and obtain a reasonable result, since it tolerated the spring rains and burning heat and stormy summer rainfall in such areas as the terrefort toulousain. Maize grains provided food for man and beast, while its husks could be used for firing and its stems for fodder. For all these reasons, maize growing in the south-west had acquired almost 'mania' proportions by the 1830s, with agricultural inspectors complaining not only that it had been grown so intensively and consistently in some areas that nutrients could not be replaced and the soil had become exhausted but also that it was being planted in environments that would never allow good yields to be produced.[15] In addition, maize growing required substantial inputs of labour and allegedly

caused tenants to neglect other work on their holdings. Landowners complained for other reasons, since tenants and sharecroppers preferred to grow maize rather than wheat and there
were fears that the new subsistence crop would take precedence
over the noble grain that figured so importantly in regional
trade. Faucher has even made the case for maize producing an
'agricultural revolution' in Aquitaine during the eighteenth
century but further innovations were not forthcoming and livestock farming was particularly neglected so that the peak of
agricultural intensification had been passed by the early
nineteenth century. [16]

At the time of the Statistique potatoes occupied 921,977 ha or
3.82 per cent of French ploughland. These figures in no way do
justice to the importance of the tuber, which accounted for 7.58
per cent of the total value of field crops (Table 5.1) and was
grown in varying degrees throughout the country (Figure 6.3a).
Lullin de Châteauvieux estimated that the potato would be
accepted in seven-eighths of France, in fact everywhere except
the Midi méditerranéen which was too hot and dry. [17] Seven core
areas were particularly important for potato growing in the
1830s. Alsace and the Vosges formed the most extensive of
these, with seven arrondissements having more than 15 per cent
of their ploughland under potatoes and proportions of between
19 and 21 per cent being recorded around Strasbourg,
Schelestadt and Weissembourg (Bas-Rhin). In some parts of the
north-east small farmers had abandoned traditional triennial
rotations in favour of a simple alternation of potatoes and rye
(or wheat or barley). As early as the 1780s Arthur Young had
noted 'more potatoes' in Lorraine 'than I have seen anywhere in
France' and the crop was particularly important in
mountainous, wooded and viticultural areas where cereals were
short. [18] In the 1830s, an even larger percentage was recorded in
the central Pyrenees, where 24 per cent of ploughland around
Saint-Girons was under potatoes. Other core areas included the
immediate environs of Paris, Upper Maine, Limousin and
Périgord, the eastern margins of the Massif Central, such as
Yssingeaux arrondissement (Haute-Loire) where the surface
under potatoes was estimated to be almost as great as that under
cereals, and Léon where in 1830 potatoes were reported to have
been grown in fields for a dozen years. [19]

The Midi and sections of the Paris Basin devoted little land to
the crop. In the former region this was because of climatic conditions and the prodigious success of maize but reasons are less
clear for the unpopularity of potatoes in the Paris Basin. It may
well have been because cereal farming was so successful and
population densities were not high away from the capital. Certainly the inhabitants of Beauce and Brie were growing only small

Figure 6.3: (a) Potatoes
and (b) Artificial
Meadows as
Percentage of Arable
Land-use

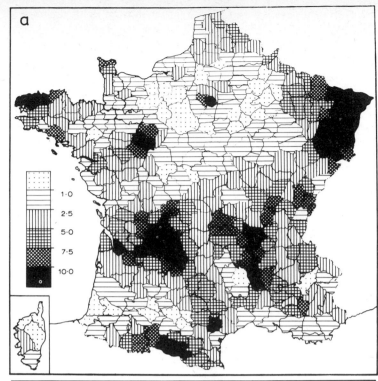

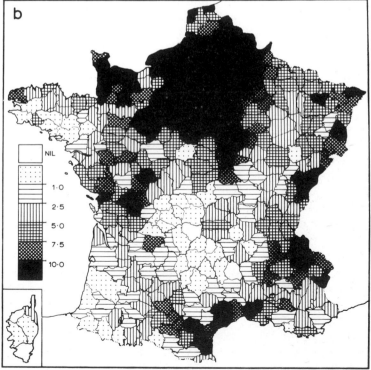

quantities of potatoes in gardens rather than in fields during the 1830s.[20] They continued to believe that potatoes were only suitable for feeding to livestock in spite of the efforts of the *Société d'Agriculture de Paris* to dispel that prejudice. By contrast, the crop was accepted most readily in pays where bread was normally a luxury. Environmental constraints on potato growing were far fewer than for maize. The tuber grew at considerable heights, in shaded areas, in poor soils and in many locations that were far from ideal for grain. Yet it was popularly believed that the potato exhausted the soil and diminished the yields of cereal crops that would follow. In addition, many farmers lacked storage areas that were free from frost, heat and damp and this contributed to the crop's lack of popularity. Finally, the nature of the diffusion process must be evoked to help account for the pattern shown in Figure 6.3a.

The origins of potato production in France are shrouded in obscurity but *truffes* (potatoes?) were recorded as being grown in Vivarais as early as 1540.[21] Certainly Arthur Young mentioned Vivarais and the whole of Upper Languedoc as one of the most important potato-growing areas in the 1780s. Yields were higher than those from cereals, fallows were reduced and for once these pays generated a food surplus that could be sold in the Rhône valley. References dating from the late-sixteenth century also relate to the Rhône valley, Alsace and the botanical garden of Duke Frédéric at Montbéliard but it is likely that many of these early examples described experiments with widespread cultivation in these areas not occurring until much later.[22]

Potatoes were, however, being grown as a field crop in western Flanders as early as 1670 but this was beyond French territory.[23] It was the German Rhinelands, and Swabia in particular, that formed the centre from which the crop was diffused into north-eastern France during the seventeenth and especially eighteenth centuries. A second but poorly known focus of early cultivation was in Dauphiné. Potatoes were being eaten in small quantities during the eighteenth century in other impoverished regions but they were shunned by most folk. This prejudice was held almost nationwide since it was popularly believed that potato eating would cause disease, with leprosy being quoted in many versions of the story.[24]

Two conditions, one administrative and the other 'natural', served to dispel long-established opposition. Intendants, ecclesiastics and agricultural societies met with varying degrees of success as they tried to popularise the potato. In 1771 Antoine-Augustin Parmentier, who had been saved from starvation in Prussia during the Seven Years' War (1756-63) by eating potatoes, wrote his prize-winning essay for the *Académie de Besançon* in which he mentioned several crops that might reduce the calamities resulting from grain shortages but paid particular attention to the virtues of the potato.[25] He continued his efforts during the Revolutionary years with the full support of the National Convention and in some départements these met with a measure of success.

None the less opposition remained strong in many pays until cereal harvests failed and prices rose. Popular attitudes began to change in parts of Alsace following the poor grain harvests of 1769-71, whilst later cereal shortages encouraged potato cultivation in Upper Poitou (1788-9), Vendée (1811), Quercy (1812), Nivernais (1812, 1816), Moselle (1817-8) and the pays charentais (1816). Trends of adoption of this kind occurred in many other parts of France and the surface under potatoes rose from 560,000 ha in 1817 to over 900,000 ha in the mid-1830s. In many areas where the potato was accepted not only were fields and gardens planted up but also the boundaries of properties, empty spaces among vines, and everywhere that the plough could not penetrate. Pockets of resistance remained, in particular in Lower Normandy where many inhabitants of Calvados and Manche refused to accept the potato in spite of medical arguments in favour of it as a rescue crop from food shortages.[26] In the relatively sparsely populated Paris Basin potatoes were only grown in gardens and farmers complained that the plant was not really suited to their soils. In other areas shortage of manure was quoted as the main reason why the crop failed to spread from gardens into arable fields, while in parts of the middle valley of the Rhône moisture was quite insufficient for potatoes to grow successfully.[27] Environmental arguments together with competition from maize explained why the potato remained essentially a mountain crop in the south-west and why it made little progress in Corsica but could not compete with chestnuts as a source of carbohydrate.[28] Yet for upland zones and areas with high population densities the potato provided an immediate addition to the human diet and also enabled it to be enriched in an indirect way since large numbers of pigs were fed on potatoes. It was as if the potato had produced a complete revolution in food supplies in the Ardennes, and the impact of the potato in the Rouergue was likened by de Rodat in Le Cultivateur aveyronnais to that of gunpowder in military history.[29]

Maize and potatoes were both grown predominantly to augment the human diet, with the feed they provided for livestock assuming an important but distinctly ancillary role. This was not the case for the range of fodder crops, known collectively as artificial meadows, which occupied previously fallow sections of ploughland in many parts of France but, in other cases, were also grown in specially fertilised small plots. Sainfoin, lucerne and two types of clover were mentioned most commonly, each with distinctive environmental tolerances and histories of diffusion. It is not possible to recognise these crops individually in the Statistique but together they occupied 1,576,547 ha or 6.54 per cent of French ploughland. During the July Monarchy they were grown most extensively in the Paris Basin and Normandy, occupying more than 15 per cent of the arable surface in three groups of arrondissements (Figure 6.3b). The first involved Seine-Inférieure plus the neighbouring arrondissement of Beauvais, with between 15 and 19 per cent of each area being under these crops. The second covered the arrondissements of Coutances (22 per cent), Caen (18 per cent) and Avranches (16 per cent); and the third comprised five arrondissements in the Ile-de-France (15-18 per cent) between Versailles and Meaux. These were not important areas for potato growing and it may well be that fodder crops usurped the role of occupying formerly fallow land that potatoes were performing in other parts of the northern third of France.

Artificial meadows did not directly provide food for human consumption and they therefore had limited appeal in largely subsistent regions where increasing the amount of carbohydrate in the human diet formed the overwhelming objective of agricultural production. They did, however, offer substantial advantages in pays that were more open to commercial forces and where traditional sources of animal fodder were slight. By definition, they enabled more livestock to be kept and more meat and dairy goods to become available for local consumption or for despatch to demand centres. More manure could be returned to the soil and this, in turn, raised cereal yields and afforded a second aspect of advance. Attractive though these opportunities would be to farmers who operated within or on the margins of commercial farming systems they would be of little or no direct interest to semi-subsistent producers whose lives were never far removed from the threat of starvation should local crops fail and their largely vegetarian diets be rendered even more slender. Artificial meadows were viewed as luxury crops by many peasants and cultivating them involved much greater adjustment than growing either maize or potatoes. Considerations such as these, as well as a range of environmental issues, may help elucidate the pattern shown on Figure 6.3b and, in particular, the virtual absence of artificial meadows from much of the Massif Central, the extreme south-west and southern Brittany. In any case, ploughland was not the leading land-use component in any of these areas.

Clover was inserted most readily into rotations since it remained in the soil for only one season. Cultivation of red clover advanced southwards from Flanders, where it was being grown as early as the fifteenth century.[30] Certainly the Flemish environment was well suited to it since soils were deep and clayey but not tenacious or excessively wet, and the temperate climate under cloud-covered skies was considered ideal. Not until the eighteenth century did red clover start to be accepted in French territory, especially in regions that were relatively cool and moist. Lime-rich soils were particularly suitable and liming prior to sowing the crop was advocated by many agronomists, including Mathieu De Dombasle. Thus, early in the eighteenth century clover was being sown in the limons of the calcareous Pays de Caux, where it occupied sections of land in the final year of the triennial rotation, but it was not until the early nineteenth century that the crop acquired widespread popularity in Upper Normandy.[31] Red clover may have appeared in Lower Alsace as early as the sixteenth century and after 1760 it became the most widely cultivated fodder crop in that region, except in the Ried where it was surpassed by lucerne and sainfoin.[32] During the eighteenth century clover was also being cultivated in Chalosse, Ille-et-Vilaine, the bocage of Poitou, the environs of Rodez and, of course, inner parts of the Paris Basin.

Greater individualism in agrarian affairs since the Revolution in theory reduced communal constraints on cultivation and enabled farmers to sow clover if they so wished; by contrast, the backward nature of sharecropping prevented experimentation with artificial meadows in Berry and elsewhere in middle France, and short leases in many areas of tenant farming also provided major obstacles.[33] Formidable hindrances also existed in areas of openfield where property was extremely fragmented and rights of vaine pâture and parcours remained in operation. Certainly this was a major complaint in Meurthe where only small areas were enclosed for growing vegetables, vines and hemp near settlements and the remainder of the cultivated landscape was one of openfield.[34] The examples of a few rich and educated people and of German-speaking Anabaptist farmers, who had planted small enclosures with clover before 1789, were emulated much more widely early in the new century.[35] Almost identical pioneering activity was reported in neighbouring Moselle, where fodder requisitions during the Revolution had forced many cultivators to experiment with red clover in the open fields even though those patches were subject to invasion by sheep and horses. Unlike red clover, trèfle incarnat was grown much less extensively. It had been known in Roussillon and Provence in the eighteenth century but was only gradually extending northwards in the early years of the nineteenth century.

Red clover was rarely cultivated in the southern half of France, and it was believed that the soil was not good enough but it was quite adequate for sainfoin (or esparcette) which grew spontaneously in poorer southern environments and reached the peak of its cultivated perfection in Languedoc and Provence.[36] Sainfoin grew successfully in soils that were low in humus, be they acid or alkaline, clayey or sandy, but was not suited to granitic or very humid soils. The crop was being cultivated in Dauphiné as early as the sixteenth century and flourished on hillslopes in Vaucluse before other types of fodder became known in Provence. It provided a valuable complement to clover and by the 1830s had been introduced successfully in a variety of locations with calcareous soils, including Champagne, the Charentes, and the Causses and Ségalas of Aveyron.

Lucerne was the fourth component in the range of artificial meadows. Unlike sainfoin it required deep, rich and permeable soils with good humus content and was acclaimed for permitting a veritable 'agricultural revolution' around Paris.[37] It did not grow successfully in tenacious soils, since it was prevented from extending its long tap roots. Lucerne flourished on alluvia in southern valleys such as the middle and lower Rhône and the environs of Grasse and Toulouse. Low-lying meadows in the Limagnes and around Valence were even ploughed up at this time and their fertile soils devoted to artificial meadows. Lucerne had been known in Alsace in the sixteenth century and in the middle Rhône in the seventeenth century but its cultivation did not extend substantially until the late-eighteenth and early-nineteenth centuries.[38] It occupied the soil for several seasons, and thus required rotations to be changed significantly if it was to be grown as a field crop. Lucerne also needed heavy manuring which was often only possible on progressive estates with their own livestock as, for example, on the plain of the Saône. These facts account for something of the ambivalence with which lucerne was regarded.

During the July Monarchy each of the fodder crops was growing in popularity, although there were important spatial variations in the trend. John Birkbeck's observation in 1814 that 'the *prairies artificielles*. . . of which so much is said by amateurs, are like specks of green in a desert' was no longer true twenty years later in Normandy and the Paris Basin at least.[39] Falling cereal prices in the 1830s encouraged farmers in Brie to devote more land to fodder crops and hence to livestock husbandry.[40] By contrast, cultivation of artificial meadows around Béthune (Pas-de-Calais) was already being overtaken by flax, colza, oil crops and beet.[41] Agronomists in less progressive parts of France continued to look to artificial meadows to improve their farming systems, and would surely have echoed Monteil's words in Aveyron: 'if ever this new form of cultivation were to find general favour in the département it would be the splendour of its agriculture'.[42]

The Established Cereals

The greater part of French ploughland was occupied by established cereals which were either gaining or losing importance during the first half of the nineteenth century. As the noble grain, figuring prominently in urban diets and most open to trade, wheat was cultivated more widely than ever before, while inferior crops, like rye, buckwheat and maslin, were retreating. Thus, for example, wheat cultivation expanded in Vexin and the Pays de Caux to eliminate maslin as fodder crops were grown, more stock were kept and more manure became available. [43] However maslin and rye were being grown more widely in some areas in the half century after the Revolution since many newly reclaimed environments were not suited to wheat but only to rustic crops.

Wheat was grown in every arrondissement in the country and formed the leading element in rotations in many regions. It covered 5,386,786 ha (23.17 per cent of French ploughland) and its production represented a crucial activity in the majority of rural communities. Harsh climates and poor soils however reduced the significance of wheat in the cropping patterns of mountainous regions and of *mauvais pays* in the lowlands (Figure 6.4a). Wheat was, in fact, grown in surprisingly difficult environments at this time of population pressure, with harvests being gathered from sunny slopes at 1,000 m in Ariège, but as a general rule, other cereals proved more rewarding in environments of moderate as well as extreme difficulty. [44]

Wheat flourished best in fertile, clayey soils, but various strains were grown in numerous types of rotation and hence it was able to occupy extensive areas of ploughland throughout France. With the exception of the Inspectors of Agriculture, few contemporary writers identified the strains that were sown and there was no mention of them in the Statistique. The wheat varieties of Languedoc, Provence and Dauphiné were reported to be superior and so cultivators often turned to southern regions when they purchased fresh seed supplies. [45] Unfortunately this kind of effort was made all too rarely in many areas. Seeding intensities and yields varied greatly between regions and hence pays with large amounts of land devoted to wheat were not necessarily the most efficient producers. Twelve arrondissements had more than double the national average of their ploughland under wheat. Eight were in the basin of Aquitaine, which reflected the role of wheat in biennial rotations, and the others were scattered widely (Nantua 70 per cent, Brignoles 57 per cent, Hazebrouck 53 per cent, Pont-Audemer 47 per cent) (Figure 6.4a). These arrondissements formed the cores of six major areas of wheat growing: the Garonne valley, Mediter-

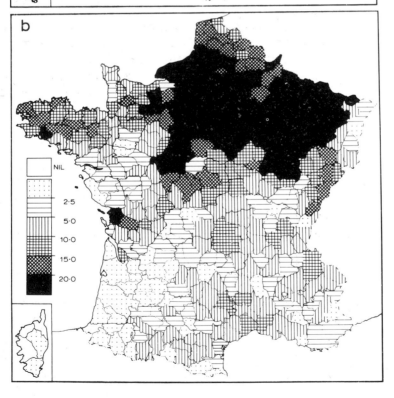

Figure 6.4: (a) Wheat and (b) Oats as Percentage of Arable Land-use

ranean France, Normandy, Nord, the Lower Loire, and eastern France from Alsace and Lorraine to the middle Rhône (with interruptions in the Vosges and the Dombes). The great granaries of Beauce and Brie had rather more modest proportions of their extensive ploughlands under wheat. Their real importance was with regard to productivity and trade rather than to area.

As well as forming the second element in the traditional triennial rotation of northern France, oats grew successfully in a wide range of environmental conditions and this fact helps explain why only four arrondissements (in Corsica and Basses-Pyrénées) reported an absence of the crop in the 1830s (Figure 6.4b). Extremes of cold, heat and drought reduced success but oats flourished on many types of soil. Even if summer heat was insufficient to mature the grain the crop would still be useful since green oats provided an acceptable source of fodder. In any case, the real importance of oats was as feed for horses, although as pottage it did enter into the human diet in Picardy, Normandy and Brittany. In many areas oats formed a resource of poorer soils, but the role of oats in the triennial rotation formed the best explanation for its distribution and this is underscored by the fact that no fewer than 33 arrondissements in central and eastern sections of the Paris Basin had between 25 and 31 per cent of their ploughland under oats. It responded well in virtually all types of terrain and microclimate in these areas and required relatively little attention, unlike the ploughing and manuring that had to precede the sowing of wheat in the first year of the triennial rotation.

The distribution of rye production formed an excellent illustration of acid soils and generally harsh environments, with the whole of the Massif Central forming an extensive ryeland and smaller examples being found in the Landes, Morbihan, Champagne, the Vosges and the Sologne (Figure 6.5a). Rye replaced wheat as the leading crop in rotation in such areas of difficulty and covered no less than 2,577,253 ha, namely 13.34 per cent of French ploughland. Only in Basses-Pyrénées (maize country par excellence) was no rye recorded in the Statistique. At the other extreme, five arrondissements in Cantal, Haute-Loire, Creuse and Corrèze had between 45 and 50 per cent of their ploughland under rye which must have been grown in a biennial rotation. Lullin de Châteauvieux argued that much French ploughland belonged more appropriately to the realm of rye rather than to that of wheat, and a quarter of a century earlier Taylor had insisted that 'rye is common in France. There is more of it gathered than of wheat, the grounds proper for it being much more extensive'.[46] He was wrong in fact but his overestimation at the national scale did not diminish the

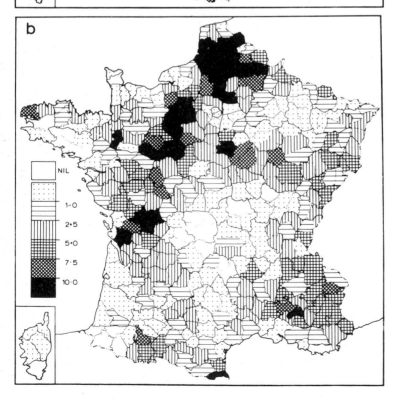

Figure 6.5: (a) Rye and (b) Maslin as Percentage of Arable Land-use

significance of rye as a vital bread grain in many of the more isolated and impoverished pays during the July Monarchy.

Nevertheless, conflicting trends were in operation. Rye growing declined as wheat cultivation extended in areas such as Vendée, Maine-et-Loire and the plateau of Langres; while, at the same time, fresh moorland was being cleared for rye in environments as precarious as 1,300 m in the high Auvergne and 1,500 m in the central Pyrenees.[47] As a winter cereal it was sown in autumn and was robust enough to survive long periods of snow cover but it was at the mercy of spring frosts, droughts or excesses of rainfall; yields were distressingly low for some years even in the Ségalas.[48] As well as providing grain for black bread, rye yielded valuable quantities of straw for thatching and for making domestic articles. In northern France it was usually produced as a petite céréale in the second year of the triennial rotation and was appreciated more as a source of fodder and straw than as a bread grain. But there were exceptions, with rye being the leading bread grain on stretches of poor land in Champagne, Burgundy, and Lorraine. In fact, the significance of rye in peasant diets in northern France should not be underestimated since it allowed wheat to be sold as a cash crop rather than being consumed locally.

Maslin (méteil) was a mixture of varying proportions of wheat and rye which was grown on soils that were not of the highest quality and where the robustness of rye was appreciated. Terminology varied from region to region and the mixture was known as conseigle in the Limagnes and molser in Alsace, where two-thirds wheat was normally sown to one-third rye. The ratio was varied to suit local conditions and differing mixtures were known as champart, méteil mitoyen as well as méteil in Eure-et-Loir.[49] Maslin was grown widely on moderate soils in the Paris Basin and in the pays charentais where it occupied the first year of the triennial rotation.[50] This fact helps account for some of the middling proportions under wheat that were recorded in Figure 6.4a. Seven arrondissements in Oise, Somme and Pas-de-Calais had between 16 and 23 per cent of their arable under maslin, by comparison with a national average of 3.78 per cent (Figure 6.5b). Contemporary opinions were divided on the desirability of sowing a mixture. In the Lyonnais and Beaujolais it was argued that maslin gave better yields than either wheat or maize in isolation but in Aveyron there were practical difficulties of harvesting two cereals that ripened at differing rates.[51] Certainly less land was devoted to maslin during the July Monarchy than in earlier times.

Buckwheat too had been grown more extensively in the past but by the 1830s a mere 2.7 per cent of French ploughland was devoted to it. It declined in coastal areas of Brittany because

seaweed, shells and sand had been applied to the soil more intensively after 1800 in order to reduce its acidity and allow wheat to
be grown.[52] Yet it remained important in the interior of
Armorica, where the population had a taste for it, in Limousin,
the middle Rhône, Ariège and the Sologne, with no less than
39.5 per cent of ploughland around Ploërmel devoted to it and
37.6 per cent around Redon. Buckwheat had been introduced
into north-western France in the late-fifteenth and sixteenth centuries, and flourished rapidly since it was initially exempt from
tithe, and was mentioned in the central Pyrenees and the Bresse
during the seventeenth century.

At least five characteristics had favoured its cultivation but
these appear to have declined in importance in the 1830s. First,
buckwheat had a very short growing season and could be sown,
for example, in the middle Rhône as late as July or August and
would yield a crop in about three months' time toward the end
of October. In less clement environments, such as the Massif
Central, its growing season was usually about four months.
Second, it was suited to most kinds of light soil and formed a
typical crop in 'poor' farming areas, needing a high degree of
humidity during the growing season. Damp summers that
proved disastrous for other cereals were therefore favourable for
buckwheat which represented 'riches next to misery, abundance
born of poverty'.[53] Nevertheless, it was still a hazardous crop
being very vulnerable to hot dry winds and precocious frosts
and this contributed to its declining popularity. Third, the short
growing season and the fact that it did not require heavy applications of manure meant that buckwheat could be sown on soil
from which other cereals such as rye had been harvested
already. Fourth, when ripe it yielded food for man and beast but
if it failed to mature it could be fed green to livestock. Finally, it
provided a subsistence crop in some wheat-growing environments and enabled the noble crop to be sold to urban consumers
or transported to other regions. This was most notably the case
in Brittany.[54] In other parts of France buckwheat was given
green to livestock, was dried for fertiliser, or its grains were fed
to poultry.

Barley covered only 4.93 per cent of the nation's ploughland
but rose to importance in Alsace, the Pays de Gex, Manche and
parts of Corsica, with each of these areas containing arrondissements with between 20 and 31 per cent of their arable under this
crop. Barley flourished on the calcareous soils of the Upper and
Lower Jurassic formations on the eastern and southern fringes of
the Paris Basin but occupied less than 20 per cent of the arable
land in each of these areas of middle France. The grain was
consumed in breweries, tanneries, stables and poultry yards and
the crop was cut green to feed to horses. It also provided a bread

grain in Alsace and the Jura, where it was grown together with or in the place of oats in the second year of the triennial rotation. [55] Its short vegetative cycle made it attractive in mountain areas such as the Pyrenees and the interior of Corsica but its role as a human foodstuff was declining. [56]

The final cereal listed in the Statistique was a group of pulses referred to as *légumes secs* which covered only 1.23 per cent of French ploughland and was entirely absent from some regions. According to Picot de la Peyrouse they were the 'grain of the poor' in the south-west, being sown 'everywhere among the maize, either in every third row or in alternative rows'. [57] Indeed between six and eight per cent of the ploughland surface of six arrondissements in Charente, Gironde and Tarn-et-Garonne was devoted to pulses, but it was in Flanders and Artois that they rose to peak importance, occupying as much as 13.2 per cent of the arable land around Montreuil. They also formed a highly distinctive component of cropping patterns in the Roumois and Lieuvin where biennial rotations were practised. [58]

Crop Combinations

The diversity of crops and fallows that composed the total arable surface may be reduced to more comprehensible dimensions and synthesised to produce a single map according to the crop-combination technique. Arrondissements which approximate most closely to the two-element model might be expected to be predominantly southern in location, reflecting the survival of modified forms of biennial rotation (Figure 6.6a). Certainly the two-element arrondissements of Provence and Aquitaine conform with this interpretation but the presence of such combinations in the Massif Central reflects the practice of long fallowing. Arrondissements with three-element combinations represent examples of well-preserved triennial rotations or of eroded biennial rotations to which new elements had been added. The predominantly southern distribution of three-element areas, namely in regions with old-established biennial rotations, would suggest that the latter case is probably more plausible. Most four-element areas were located in the Paris Basin and this implies that they represent triennial rotations that have been modified by the addition of new elements, which would probably include artificial meadows. Finally, one may recognise arrondissements with complicated polycultural patterns comprising five, six or even seven elements.

In fact, Figure 6.6a poses serious problems of interpretation. For example, two-element areas involved unquestionably harsh environments in the cool Massif Central but also hotter and more arid conditions in the Midi where the biennial rotation remained the reasonable ecological response to summer drought.

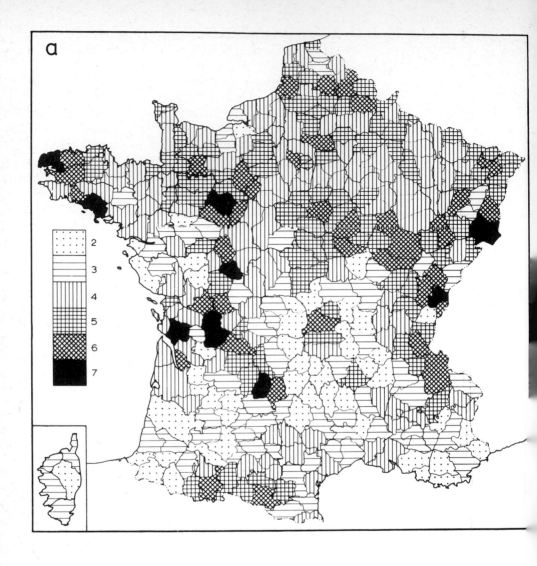

a

. . .	2
───	3
‖‖‖	4
⊞⊞⊞	5
▨▨▨	6
■■■	7

The presence of many elements in a crop combination might indicate that an area was 'open' to the cultivation of many new cash crops but might equally well reflect that it was relatively 'closed' and needed to provide a range of commodities for local subsistence. Caution should be exercised in interpreting the number of elements indicated on Figure 6.6a. In some instances an element might correspond to a season, thus a three-element model might represent a triennial rotation, but in many cases the relationship was by no means so straightforward. In addition, there were important differences in the internal composition of crop combinations displaying identical numbers of elements. For example, both Upper Normandy and Champagne displayed

Figure 6.6: (a) Number of Elements and (b) Crop Combinations

104

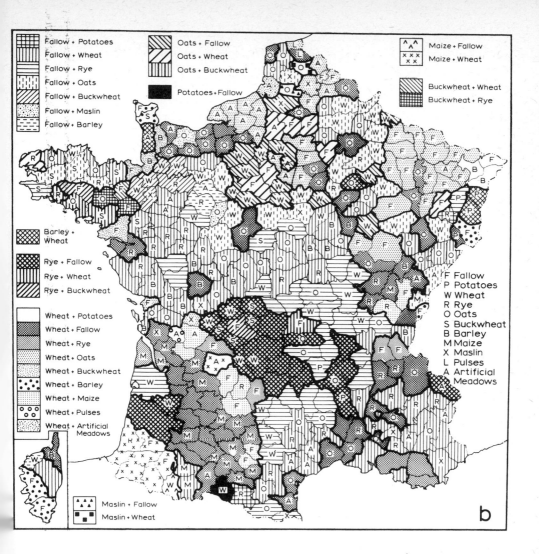

Fallow + Potatoes
Fallow + Wheat
Fallow + Rye
Fallow + Oats
Fallow + Buckwheat
Fallow + Maslin
Fallow + Barley

Oats + Fallow
Oats + Wheat
Oats + Buckwheat

Potatoes + Fallow

Maize + Fallow
Maize + Wheat

Buckwheat + Wheat
Buckwheat + Rye

Barley + Wheat

Rye + Fallow
Rye + Wheat
Rye + Buckwheat

Wheat + Potatoes
Wheat + Fallow
Wheat + Rye
Wheat + Oats
Wheat + Buckwheat
Wheat + Barley
Wheat + Maize
Wheat + Pulses
Wheat + Artificial Meadows

F Fallow
P Potatoes
W Wheat
R Rye
O Oats
S Buckwheat
B Barley
M Maize
X Maslin
L Pulses
A Artificial Meadows

Maslin + Fallow
Maslin + Wheat

b

four-element combinations which reflected eroded triennial rotations, but in the former case the cropping pattern had been enhanced by the inclusion of large amounts of artificial meadows and a parallel contraction of bare fallows, while in the latter case fallows remained the most extensive element in the combination, since artificial meadows had made very little impact.

Such subtleties of composition are elucidated in Figure 6.6b which summarises the distribution of each of the crops. Wheat was the leading field crop in eight regions. The most extensive of these was Aquitaine, extending from the Pyrenees as far north as Ruffec in Charente. The normal combination was 'wheat and fallow' which evoked the traditional biennial rotation with maize sometimes entering as the third element and a few arrondissements having maize or rye as the second most important element. Lorraine, northern Alsace and Burgundy formed the second large region dominated by wheat, with 'wheat and oats and fallow' being the normal combination and reflecting the triennial rotation. 'Wheat and potatoes' characterised northern Alsace, and maize entered in second or third place in the Bresse. Dauphiné formed a compact block of wheat territory, usually with rye in second or third position. Wheat dominated crop combinations in only ten arrondissements in the Mediterranean south and these were set in an enormous stretch of 'fallow and wheat' territory which covered the remainder of Languedoc and Provence and really just represented a less intensive system of wheat growing in accordance with the biennial rotation. The *pays nantais* was a small outlier of wheat-dominated territory, with buckwheat entering combinations in second position on the north bank of the Loire. Wheat dominated crop combinations in the greater part of Normandy from Saint Malo to Dieppe but there were interesting local variations in the second and third elements: 'wheat and oats and artificial meadows' in a solid block of Seine-Inférieure; 'wheat and fallow and oats' for much of Calvados and Eure; 'wheat and artificial meadows and oats' in the Campagne de Caen; 'wheat and barley' in Cotentin; and 'wheat and buckwheat' on the margins of Armorica. Wheat was also the leading crop in Nord and neighbouring districts just to the south, with oats normally being in second place. Similar combinations dominated in a horseshoe-shaped group of arrondissements to the north, east and south of Paris. Indeed, oats formed the leading crop in Beauce and in much of Picardy so that when arrondissements characterised by 'oats and fallow and wheat' and 'oats and wheat and fallow' are included the bulk of the intervening area between Normandy, Brie and Nord is accounted for. In this way it is possible to start to assemble the great region of triennial rotations across northern France. The picture becomes even more convincing when the 'fallow and oats and wheat' and 'fallow and wheat and oats' combinations in Champagne, Burgundy, Franche-Comté and arrondissements on the southern margins of the Paris Basin are considered.

Rye dominated a range of environments in the Massif Central, the Landes and high Alps. Fallow normally formed the second element, with various crops (wheat, buckwheat, potatoes, artificial meadows) in third position depending on conditions in adjacent areas. Each of the remaining crops emerged in leading position in only small areas. Pulses failed to occur in leading position but appeared in second or third place in two arrondissements in Flanders and around Cognac. The remainder of France displayed combinations in which fallow was the most extensive component. Middle France formed the intersection of an irregular cross-shaped assemblage of combinations, from Lower Brittany through the marchlands of Armorica and the northern margins of the Massif Central southwards to Provence, and from the eastern Pyrenees through southern and eastern sections of the Massif Central into Champagne, with an outlier in Franche-Comté and the Vosges. The crops in second position in the arrondissements of the cross need to be examined carefully since these fallow-dominated areas simply represent relaxations on the margins of regions that have been discussed already. The arrondissements of the cross were those where agricultural intensification might well take place by the progressive elimination of fallows from existing cultivated fields, and these zones also contained large expanses of uncultivated land and so there was also scope for défrichement.

Notes

1. M. Bloch, *Les Caractères originaux de l'histoire rurale française* (Paris, 1931); R. Dion, *Essai sur la formation du paysage rural français* (Tours, 1934).

2. D. Faucher, 'L'assolement triennal en France', *ER*, vol. 1 (1961), pp. 7-18.

3. F. Sigaut, 'Pour une cartographie des assolements en France au début du XIXe siècle', *AESC*, vol. 31 (1976), pp. 631-43.

4. Inspecteurs de l'Agriculture, *Agriculture française: Haute-Garonne* (Paris, 1843), p. 101.

5. M. Birkbeck, *Notes on a Journey through France* (London, 1814), p. 101.

6. M. de Chambray, 'De l'agriculture et de l'industrie dans la province de Nivernais', *AAF*, vol. 13 (1834), p. 8.

7. AD Meurthe, 7 M 117. *Statistique*: généralités, 1815-50.

8. E. Robert, 'Considérations sur les assolements et l'abolition des jachères en Provence', *APAPER*, vol. 13 (1840), pp. 105-16.

9. H. de Villeneuve and E. Robert, 'Revue agricole de la Provence', *APAPER*, vol. 12 (1839), p. 40.

10. H. Laure, *Guide des cultivateurs du midi de la France, de la Corse et de l'Algérie* (Toulon, 1854), p. 268; M. Bosc, 'Description générale et statistique du département de l'Aude', *AAF*, vol. 6 (1819), p. 387.

11. A. Young, *Travels during the Years 1787, 1788 and 1789* (London, 1792), p. 363; A. Armengaud, *Les Populations de l'Est aquitain* (Paris, 1961), p. 109.

12. B.H. Slicher Van Bath, *The Agrarian History of Western Europe A.D. 500-1850* (London, 1963).

13. D. Faucher, 'Le maïs en France', *AG*, vol. 40 (1931), pp. 113-21; J. Klatzmann, *La Localisation des cultures et des productions animales en France* (Paris, 1955), pp. 271-2.

14. P.M. Hohenberg, 'Maize in French agriculture', *JEEH*, vol. 6 (1977), pp. 63-101.

15. Inspecteurs de l'Agriculture, *Agriculture française: Tarn* (Paris, 1845), p. 260; Inspecteurs de l'Agriculture, *Agriculture française: Aude* (Paris, 1847), p. 193.

16. D. Faucher, 'La révolution agricole du XVIII-XIXe siècle' in *La Vie Rurale vue par un géographe* (Toulouse, 1962), pp. 143-72.

17. F. Lullin de Châteauvieux, *Voyages agronomiques en France* (Paris, 1843), p. 334.

18. Young, *Travels*, p. 77.

19. AD Finistère 100 J 791. Administration générale et économie. L'état de l'agriculture dans le Finistère par Richard (1830).

20. J. Doublet de Boisthibault, *La France: description géographique, statistique et topographique: Eure-et-Loir* (Paris, 1836), p. 142; A. Rayer, *Etude sur l'économie rurale du département de Seine-et-Marne* (Paris, 1895), p. 70.

21. P. Bozon, *La Vie rurale en Vivarais* (Clermont-Ferrand, 1961), p. 81.

22. E. Tisserand and L. Lefebvre, *Étude sur l'économie rurale de l'Alsace* (Paris, 1869), p. 47.

23. M. Morineau, 'La pomme de terre au XVIIIE siècle', *AESC*, vol. 25 (1970), pp. 1767-85.

24. E. Roze, *Histoire de la pomme de terre* (Paris, 1898).

25. M. Hétarit, *Histoire générale de la pomme de terre* (Troyes, 1849), p. 11.

26. G. Désert, 'La culture de la pomme de terre dans le Calvados au XIXe siècle', *AN*, vol. 5 (1955), pp. 261-70.

27. D. Faucher, *Plaines et bassins du Rhône moyen* (Paris, 1927), p. 368.

28. A. Moreau de Jonnès, 'Statistique agricole de la Corse', *Le Cultivateur*, vol. 18 (1842), pp. 686-98.

29. A. Bouvart, 'Agriculture des communes forestières des Ardennes', *JAP*, vol. 6 (1842-3), pp. 98-102, 395-9; cited in R. Béteille, *Les Aveyronnais* (Poitiers, 1974), p. 29.

30. D. Faucher, *Géographie agraire: types de cultures* (Paris, 1949), p. 181.

31. J. Sion, *Les paysans de la Normandie orientale* (Paris, 1909), p. 225; A. Frémont, *L'Elevage en Normandie* (2 vols., Caen, 1967), vol. 2, p. 451.

32. N. Schwerz, *Assolements et culture de l'Alsace* (Paris, 1839), p. 217.

33. M. De Puyvallée, 'Rapport sur les moyens de concilier avec la culture par métayers l'établissement et la conservation de la culture des prairies artificielles', *Bulletin de la Société d'Agriculture du Département du Cher*, vol. 8 (1825), p. 5.

34. AD Meurthe, 7 M 117. Statistique: généralités, 1815-50.

35. M. Marquis, *Mémoire statistique du département de la Meurthe* (Paris, 1803), p. 166; M. Verronais, *Statistique historique, industrielle et commérciale du département de la Moselle* (Metz, 1844), p. 114.

36. M. Bosc, 'Description générale et statistique du département de l'Aude', *AAF*, vol. 6 (1819), p. 395.

37. Lullin de Châteauvieux, *Voyages*, p. 313.

38. Tisserand and Lefebvre, *L'Economie rurale de l'Alsace*, p. 48.

39. M. Birkbeck, *Notes on a Journey through France* (London, 1814), p. 110.

40. AD Seine-et-Marne, M 7323. 'Situation et progrès de l'agriculture, 1835'.

41. Anon., 'Précis statistique sur le canton de Cambrai', *Annuaire Statistique et Administratif du Pas-de-Calais* (1848-9), p. 38.

42. A.A. Monteil, *Description du département de l'Aveyron* (Paris, 1803), p. 30.

43. Sion, *Les Paysans*, p. 355; J. Dupâquier, 'La situation de l'agriculture dans le Vexin français fin du XVIIIe siècle et début du XIXe siècle', *ACSS*, (1964), pp. 321-45.

44. M. Chevalier, *La Vie humaine dans les Pyrénées ariégeoises* (Paris, 1956), p. 225.

45. J.N. Taylor, *Sketch of the Geography, Political Economy and Statistics of France* (Washington, 1815), p. 315.

46. Lullin de Châteauvieux, *Voyages*, p. 295; Taylor, *Sketch of France*, p. 316.

47. M. Lange, 'Rapport sur un ouvrage de Mr. Yvart', *Mémoires de la Société d'Agriculture et de Commerce de Caen*, vol. 3 (1830), p. 71; Chevalier, *Pyrénées ariégeoises*, p. 225.

48. H. Cavaillès, *La Vie pastorale et agricole dans les Pyrénées des Gaves, de l'Adour et des Nestes* (Paris, 1931), p. 176; A. Meynier, *A Travers le Massif Central: Ségala, Levezou, Châtaigneraie* (Aurillac, 1931), p. 110.

49. J. Doublet de Boisthibault, *La France: description géographique, statistique et topographique: Eure-et-Loir* (Paris, 1836), p. 36.

50. P. Goubert, 'Les techniques agricoles dans les pays picards aux XVIIe et XVIIIe siècle', *RHES*, vol. 35 (1957), p. 28; J.P. Moreau, *La Vie rurale dans le sud-est du bassin parisien* (Paris, 1956), p. 97; J.P. Quénot, *Statistique du département de la Charente* (Paris, 1818), p. 393.

51. G. Garrier, *Paysans du Beaujolais et du Lyonnais, 1800-1970* (2 vols., Grenoble, 1973), vol. 1, p. 169; Monteil, *Aveyron*, p. 53.

52. Inspecteurs de l'Agriculture, *Agriculture française: Côtes-du-Nord* (Paris, 1844), p. 181.

53. P. La Boulinière, *Itinéraire déscriptif et pittoresque des Hautes-Pyrénées françaises* (3 vols., Paris, 1825), vol. 3, p. 292.

54. R. Musset, *La Bretagne* (Paris, 1948), p. 79.

55. Schwerz, *Assolements d'Alsace*, p. 141; R. Lebeau, *La Vie rurale dans les montagnes du Jura méridional* (Lyons, 1955), p. 280.

56. Cavaillès, *La Vie pastorale*, p. 176; L. Moll, 'Agriculture de la Corse', *JAP*, vol. 1 (1837-8), p. 168.

57. Picot de la Peyrouse, *The Agriculture of a District in the South of France* (London, 1819), pp. 49-50.

58. L. Musset, 'Observations sur l'ancien assolement biennal du Roumois et du Lieuvin', *AN*, vol. 2 (1952), pp. 143-50.

Seeding and Productivity

The essential function of agriculture is to produce consumable commodities and hence it is necessary to turn from land use and attempt to determine spatial variations in productivity and production in the arable realm during the July Monarchy. In order to be intelligible, patterns of production must be related to patterns of consumption and to the spatial variations in commodity prices which characterised the country prior to its effective integration by the railway network. Such matters are clearly fundamental in attempting to determine the relative efficiency of agricultural activity throughout the country, however contemporaries wrote little about them and they have received only slight attention from recent scholars. Michel Morineau demonstrated the superiority of wheat and rye yields in Nord département and the broad contrast between high- and low-yielding areas on either side of a line between Les Sables-d'Olonne and Lake Geneva.[1] Admittedly wheat and rye formed the major sources of bread grain but the other cereals deserve scrutiny in order to discover if spatial variations in their productivity conformed to the patterns produced for the leading grains or whether individual crops produced their own discrete patterns in relation to the particular ecological conditions to which they were best suited.

Commodity prices fluctuated in response to the quality and quantity of harvests but, in addition, spatial variations in the relationship between supply and demand produced striking differences in their geographical arrangement before the railway allowed supplies to be moved effectively and relatively cheaply. The physiocrats in the eighteenth century had hoped that freedom of trade would cause spatial differences in wheat prices to decline if not disappear but that was not to be the case until after 1850. During the preceding hundred years the market structure remained fragmented and a fairly consistent pattern of wheat prices involved an array of cheap areas north of the Loire and expensive areas to the south, with a secondary expensive outlier in the Lower Seine.[2] Whether the prices of other commodities conformed to the pattern for wheat has not been examined by researchers in a comprehensive way. Fortunately the Statistique devoted a considerable amount of space to this matter. A sizeable proportion of cereals grown during the July Monarchy escaped the commercial economy but there is little reason to believe that the 'corrected' arrondissement averages were less than reasonable reflections of local market conditions.

Contemporary journals were virtually silent on seeding practices in the 1830s save to remark that seed was broadcast in even the most progressive areas.[3] The stated average volume of wheat applied per hectare varied enormously, from a mere 1.07 hl around Rochechouart (Haute-Vienne) to 3.75 hl around Briançon (Hautes-Alpes) (Figure 7.1a). The lowest sextile of intensity (less than 1.72 hl/ha) involved Corsica and a fairly compact block of south-western France which included environments as diverse as those of the Garonne valley, where wheat was the leading crop, and parts of Limousin, where it was of only slight importance. By contrast, seeding intensities were high in northern and eastern France, involving areas that formed an inverted 'V' shape from Finistère, through Normandy and the core of the Paris Basin to Alsace, Franche-Comté and Dauphiné, and thence to Lower Auvergne and the Alps. According to Monteil the general rule was that 'the more fertile the land, the more seed that had to be sown; the more infertile it was the less seed that was used' although some modern sources state the reverse opinion.[4] Certainly the pattern shown in Figure 7.1a did indeed conform in general terms to Monteil's proposition. High intensities applied in parts of the Alps must have related to rich, valley-bottom soils because the noble grain would not have ripened in harsher, mountainous surroundings. It would also appear that seed applications were low in areas with biennial rotations and high where triennial rotations operated. The link may well be through Faucher's assertion that triennial rotations could only operate successfully on more naturally fertile soils.[5] Intense seeding involved not only commercially open areas, such as the granaries of the Paris Basin and Dauphiné, but also rather more isolated environments in eastern and north-eastern France where wheat, none the less, was a prized crop for local consumption and in some areas also for trade.

For France as a whole 12.45 hl were being harvested from every hectare devoted to wheat in the 'average year' of the 1830s (Table 7.1). Dezeimeris remarked with concern that this was only half of what was being achieved in England and in parts of Belgium and Germany.[6] Crude yields of 6-7 hl/ha passed as good in the Midi, even where soils were fertile. Commentators were virtually unanimous that the only way to improve matters was to devote less land to wheat and more to fodder crops as a means of raising more livestock and generating more manure. Nivière argued that cereal yields in areas of grande culture might be doubled in that way.[7] Crude yields were almost entirely below 10.0 hl/ha in central and southern France, with the exception of Dauphiné, Lower Auvergne, Velay, parts of Languedoc

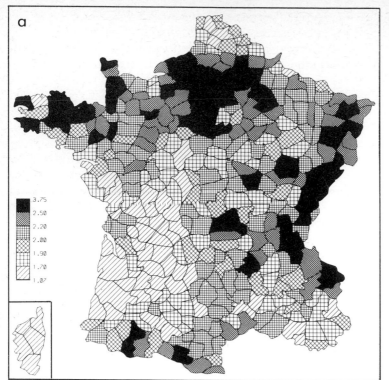

a

■	3.75
▨	2.50
▨	2.20
▨	2.00
▨	1.90
▨	1.70
	1.07

Figure 7.1: Wheat
(a) Seed Applied
(hl/ha) by Sextiles
(b) Crude Yeilds (hl/ha)
by Sextiles

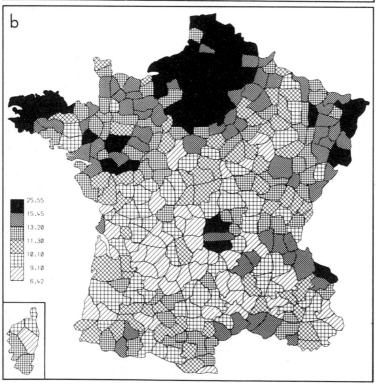

b

■	25.55
▨	15.45
▨	13.20
▨	11.30
▨	10.10
▨	9.10
	6.42

and the western Pyrenees (Figure 7.1b). North of the Loire yields were almost entirely above that amount, with the exception of parts of Champagne, Nivernais, the *pays blésois* and eastern Morbihan. Sixteen arrondissements recorded crude yields of 20.0 hl/ha or more: in a block of seven arrondissements in Nord, Pas-de-Calais, Aisne and Oise; in the environs of Paris; around Pont-Audemer at the mouth of the Seine; near Lannion and Saint-Brieuc; and in the Briançonnais. Arrondissements in the first group generally involved good soils, with rotations containing very little fallow in the extreme north of France and incorporating an important proportion of artificial meadows in the inner Paris Basin. Intensive cultivation, additional supplies of manure and very large demands from the capital and the cities of Flanders played substantial roles in fuelling agricultural systems that would produce such high yields. Coastal Brittany, with its marine fertilisers and small farms, was also able to maintain high productivity for wheat which, like vegetables, was an 'export' crop. The highest crude yield was recorded around Meaux (25.5 hl/ha) in Brie and was four times greater than the lowest yield in the country (6.4 hl/ha around Gourdan in Lot). The low wheat yields of the Massif Central were far from unexpected for this was a rye land rather than wheat country. By contrast, the Basin of Aquitaine was predominantly wheat country but even there many arrondissements registered yields below 10.0 hl/ha or even 9.0 hl/ha in the case of arrondissements in Dordogne and Lot.

The pattern of net yields was virtually identical to that for crude yields but with the volume used for seeding subtracted from the total output, which served to scale down values by

Table 7.1: Production and Prices of Cereals and Potatoes

	Total Production	Seed	Net Production	Consumption/Net Production (%)	Seed: Yield Ratio	Crude Yield (hl/ha)	Average Price (F/hl)
	('000 hl)						
Wheat	69,558	11,441	58,096	99.18	1:6.1	12.45	15.85
Oats	48,899	7,015	41,884	87.38	1:6.9	16.29	6.20
Rye	27,811	5,139	22,672	98.09	1:5.4	10.79	10.65
Barley	16,661	2,575	14,085	88.05	1:6.5	14.02	8.25
Maslin	11,829	1,932	9,897	113.26	1:6.1	12.98	12.20
Buckwheat	8,469	551	7,918	88.39	1:15.1	13.00	7.25
Maize	7,620	242	7,377	90.24	1:31.4	12.06	9.40
Potatoes	96,233	10,267	85,966	91.24	1:9.4	104.37	2.10
Spelt	136	15	120	112.25	1:8.6	28.76	5.95
Pulses	3,460	539	2,920	106.63	1:6.4	11.65	15.05

roughly 2.0 hl (Figure 7.2a). As with Figure 7.1 it is notable that the upper sextile spans a very wide range of values, in the order of 10.0 hl. Only three arrondissements had net yields of more than 20.0 hl/ha (Lille and Hazebrouck 20.3, Meaux 22.9), with the lowest being Saint-Etienne (4.6). Central and northern parts of the Paris Basin, Upper Normandy, Lower Brittany, Alsace, the Limagnes and Dauphiné formed the highest yielding areas, whilst levels of productivity in Aquitaine were moderate or even low. The regional pattern of net productivity was thus substantially different from what the pattern of land use and crop combinations might be taken to imply. In other words, areas in which wheat was the dominant user of land were not always efficient producers of that commodity.

Crude yields need to be related to seed applications in order to obtain a more meaningful index of productivity. Seed:yield ratios provide the simplest expression of this relationship (Figure 7.2b). In average terms each hectolitre of wheat sown in France in the 1830s yielded 6.1 hl. The seed:yield ratio was thus 1:6.1, precisely as Arthur Young had estimated fifty years earlier but he did emphasise that there were important differences between districts (loams 1:8, heath 1:6, mountain 1:5, gravel 1:5, various soils 1:5, stony soils 1:4).[8] Only sixty-eight arrondissements registered ratios of 1:7 or more in the 1830s and very few of these were to the south of the Loire. With the exception of Marseilles (1:9.0), Meaux (1:9.7) and Morlaix (1:10.2), all areas with ratios of 1:9.0 or higher were located in the three départements of Nord, Pas-de-Calais and Somme, with ratios of 1:11.3 to 1:11.7 around Lille, Béthune, Hazebrouck and Avesnes. Ratios for the granaries of Beauce and Brie were certainly high, as were those for northern Alsace and the Pays de Caux, but they were not in the top league. As far as wheat yields in the southern half of France were concerned, only the Limagnes, the lower Rhône, parts of the Garonne valley, Limousin, the pays charentais and Corsica recorded average or above average seed:yield ratios. A broadening but fragmented crescent of truly productive territory extended from Vendée to Alsace, with important interruptions in response to the harsher environments of Arcoët, Perche, Champagne and the Vosges. The legendary fertility of intensive farming systems was amply demonstrated in Flanders and the surrounding districts but important contrasts in productivity existed between grande and petite culture. For example, Planche estimated that wheat yields from areas of petite culture in the Midi were virtually double those generated by grande culture, and Nivière came to the same conclusion for France as a whole.[9] Unfortunately, such differences may not be identified from the Statistique.

Unlike wheat, oats was used predominantly as a feed for

Figure 7.2: Wheat
(a) Net Yield (hl/ha) by
Sextiles
(b) Seed:Yield Ratio

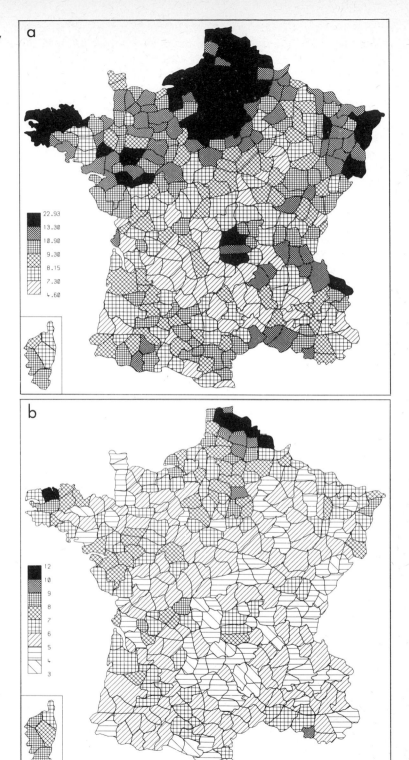

livestock but it was associated with wheat in traditional triennial rotations and, like that crop, entered into domestic trade. Indeed, the intensity of seeding of oats bore certain similarities with that for wheat. Relatively small volumes of oats were applied throughout south-western France (normally less than 1.70 hl/ha), while there were high intensities of application in harsher and more hazardous environments, where the risk of seed failure must have been great. By contrast with wheat, only four arrondissements at the core of the Paris Basin came in the top sextile for oats seeding, but much of Brittany and parts of Lower Normandy appeared in that category, with roughly 3.00 hl/ha being sown.

Even though only a small number of arrondissements in the Paris Basin experienced the highest intensity of seeding, a very much larger section of this generally fertile region appeared in the top sextile of crude yields. Areas from Le Havre to Rocroi and from Dunkirk to Corbeil produced over 20.5 hl/ha, with yields of more than 35.0 hl/ha throughout Nord and no less than 44.4 hl around Lille and 49.3 hl near Douai. Lower Brittany, the Lower Rhône and the Bresse formed detached areas with high yields. Natural and culturally-enhanced fertility, openness to commercial forces, operation of triennial rotations and the presence of a large equine population combined to produce this pattern. At the other extreme, a crude yield of only 6.4 hl was registered around Mende in Lozère. Oats was a higher yielding crop than wheat and for this reason rather modest quantities of seed were applied to the fertile ploughlands of Picardy and parts of the Ile-de-France. Absolute values of productivity are scaled down by roughly 2.0 hl when volumes of seed are subtracted. The average seed:yield ratio for oats was 1:6.9 and no fewer than 138 arrondissements recorded ratios of 1:9 or more, by comparison with only fifteen for wheat. Ratios in the intensively cultivated heart of Nord (Lille 1:19, Douai 1:21) were roughly double those for the Ile-de-France, Strasbourg, the Bresse, Vendée and the environs of Morlaix. In the face of such high productivity, the ratios of less than 1:4 in Lozère and Franche-Comté were low indeed. As was the case for wheat, the massive superiority of yields from Nord was quite unchallenged.

Such too, must be the conclusion with respect to rye, even though Nord was certainly not one of the ryelands of France. As with wheat and oats, seeding intensities were very low over much of the Basin of Aquitaine and in the Landes, with little over 1.0 hl/ha being sown. Two and a half times that quantity or even more were used in parts of Brittany, Normandy, Alsace, Franche-Comté, Dauphiné and the Pyrenees so that the pattern for rye seeding bore many similarities with that for wheat. Crude yields ranged from 5.5 hl/ha at Apt to 23.8 hl around

Châteaulin. With the exception of the Limagnes, yields in the Massif Central were below 10.0 hl and in much of central France they fell below 8.5 hl. Other areas, such as the Landes and eastern parts of the Paris Basin, which had sizeable proportions of their ploughland under rye, also recorded low yields, which declined to between 6.0 and 7.0 hl in parts of Champagne and Burgundy. Northern parts of the Paris Basin, Poitou and Lower Brittany displayed the highest levels of productivity. Subtraction of seed scaled down average rye yields by just under 2.0 hl but the general picture of net productivity remained essentially the same as those for wheat and oats with the addition of areas in Poitou and toward the mouth of the Loire.

The range of seed:yield ratios for rye resembled that for wheat but the national average was only 1:5.4. Ratios of 1:9 or greater involved only fourteen arrondissements, predominantly in Nord but also in Lower Brittany, Poitou and around Bayonne (with a peak ratio of 1:12). When ratios of 1:8 and 1:7 were included the *poitevin* focus of rye productivity emerged across the western section of middle France. Productivity was particularly low in Cantal, the plain of Forez and parts of Franche-Comté and Champagne, where ratios were only 1:3. There were, of course, very great variations from season to season for this crop of mountains and moorlands, with, for example, ratios in the Ségalas ranging from 1:10 in exceptionally good years to only 1:2 in disastrous seasons. [10]

The lower application of barley seed in south-western and central-western France and high applications in Franche-Comté and parts of the Alps roughly replicated the pattern that has been outlined already but additional foci of high intensity seeding involved Marne and Manche and sections of the Massif Central and the central Pyrenees. Over much of the Massif Central, Champagne and many areas along the eastern border more than 2.2 hl/ha were sown, with double that intensity being reached around Pontarlier, Montbéliard and Briançon. Neither Nord nor the Ile-de-France involved particularly high intensities of seeding but crude yields displayed the by-now familiar pattern with over 30.0 hl/ha being produced throughout Nord and no less than 40.2 hl around Brest. Lower Auvergne and Dauphiné formed minor concentrations of high productivity, yielding 19.0-21.0 hl, with the lowest crude yields (7.0 hl) being in Haute-Marne. Calculation of net yields hardly modified the picture (Figure 7.5a). Seed:yield ratios averaged 1:6.5, rising to 1:16 around Brest and in parts of Nord and Pas-de-Calais (Figure 7.5b). At the other extreme the ratio was only 1:3-4 in Franche-Comté, Burgundy, the Pyrenees, Nivernais, Bourbonnais, the plain of Forez and Manche, even though barley was important in crop combinations in the latter département.

Maslin was not grown in all parts of France, nor was the ratio of wheat and rye constant, hence information must be treated with caution. The essential contrasts of seeding between the south-west and the north plus many upland areas emerge again. Only 1.20 hl/ha was sown along the western fringes of the Massif Central, whilst four times that amount was used around Dôle (4.52 hl). The highest crude yields were predictably in Nord (Lille 22.4 hl) and the inner Paris Basin, with Lower Brittany (Brest 22.4 hl) and Lower Auvergne forming detached areas of high productivity. Net yields were as low as 4-6 hl in Burgundy, the southern fringes of the Massif Central and much of the south-eastern corner of France. Only fourteen arrondissements generated seed:yield ratios of 1:9 or above and, not surprisingly, eleven of these were located in Nord and Pas-de-Calais. The national average was identical to that for wheat, namely 1:6.1.

Buckwheat was characteristically a high-yielding crop and the application of seed was therefore less intense than for other crops. The highest intensities involved areas as widely distributed as Aisne, Eure, Haute-Saône, the middle Rhône and Lower Languedoc with low intensities being numerous in Lower Poitou and Anjou. Crude yields were low on the southern margins of the Massif Central, whilst the largest areas with very high yields were found in Brittany (15-20 hl/ha) and Dauphiné (22 hl/ha). However, these were substantially below the peak yield of 27.17 hl around Douai. Net yields were only marginally lower because of the high yielding nature of buckwheat with an average seed:yield ratio of 1:15.4. No fewer than fifteen arrondissements had seed:yield ratios of 1:25 or more, with the highest being in Douai, Lannion, Les Sables-d'Olonne (each 1:32) and Fontenay (1:34). The highest ratios were not in Upper Brittany or the Armorican fringes, which were the leading areas for buckwheat in terms of land use, but rather along the Armor fringe of Lower Brittany, in Limousin, Lower Poitou and especially Vendée. Environmental conditions for growing buckwheat were particularly favourable in Armorica and Limousin and this fact produced a substantially different pattern of productivity from that for other crops. The presence of Douai in joint second position may be disregarded since only 6 ha were involved.

Maize was not grown in many parts of France and, as with buckwheat, the picture is fragmented and hard to interpret. The major focus of maize growing in the Pays de l'Adour involved only low or moderate rates of seed application, with much higher intensities being used in rather more marginal environments in the Saône valley, in Dauphiné and a scatter of areas on the southern margins of the Massif Central, in Lower

Languedoc and Provence. The ideal growing conditions of the Pays de l'Adour produced some of the highest crude yields (e.g. 24.8 hl/ha around Argelès) but the top sextile was represented more prominently in the pays nantais, Dauphiné and Le Comtat. By contrast, productivity in Dordogne and Tarn was low, with net yields of 6.0 to 8.5 hl, compared with over 30.0 hl in Dauphiné. This point emerges even more clearly from seed:yield ratios which averaged 1:31.4. Ratios in four arrondissements of the extreme south-west were high (just over 1:50) but they were exceeded in Vendée and around Valence (1:65) and Avignon (1:100). Values for the terrefort toulousain (*c.* 1:40) and Dordogne and Lot (1:20 to 1:30) were appreciably lower. As contemporaries observed, maize productivity in the south-west was none the less fabulous, when compared with other crops, being five to eight times that for wheat in the terrefort toulousain and in some years being even greater.[11]

Information on the regeneration of potatoes and artificial meadows was not expressed in the same fashion as that for cereals and may not be examined in the same way. The equivalent of a seed:yield ratio for potatoes averaged 1:9.4 and ratios were high in the intensively cultivated pays of Flanders and Artois, in parts of Normandy and around Meaux, with values of 1:12 to 1:14 and very occasionally 1:16. Ratios were marginally higher still in Lower Brittany (1:13 to 1:16) and Limousin (1:12 to 1:17). In many areas where the potato was an important occupier of land, ratios were moderate or even low. For example, ratios of 1:6 to 1:8 were usual in the north-east, Upper Maine and Dordogne, rising to 1:9 to 1:10 in eastern parts of the Massif Central where productivity was only average.

The spatial arrangement of seeding practice and crop productivity varied in detail in response to the ecology of each particular crop, however three generalisations may be drawn.

1. A fundamental contrast existed between high intensity seeding in south-western France and low intensities in north-western areas. Responses to the environmental diversity of the Massif Central mean that a simple spatial gradient may not be recognised, but a contrast between the south-west and the north-east was in evidence for wheat, oats, rye and barley, which were grown throughout the country, and high intensities of maslin seeding also conformed to this general pattern.

2. Yields, whether expressed in crude or net terms or more meaningfully as seed:yield ratios, were higher in the northern third of the country with its triennial rotations, more regular supplies of fodder, higher densities of livestock and greater amounts of manure. Only maize, with specifically southern ecological requirements, formed an exception to this rule. Arrondissements in Nord and Pas-de-Calais came at the top of

the league not only for wheat and oats, as may be expected, but also for rye, barley, maslin and buckwheat, even though only small quantities of the latter crops were grown. Flemish productivity was surpassed by that of Lower Brittany as far as potatoes were concerned, and maize would not, of course, ripen in the 'far north'. But in every other respect the superiority of the ploughlands of Flanders and its surrounding pays remained unquestioned.

3. Crops were often only moderately productive in areas where they dominated crop combinations. Thus, wheat yields were only moderate or low in the south-west; rye yields were below average in the Massif Central; buckwheat yields were not at their highest in the interior of Upper Brittany; nor were the highest potato yields recorded in the north-east. In other words, measures of land use provide a very incomplete view of the significance of particular crops in agricultural production.

Production, Consumption and Prices

Spatial variations in grain productivity were of more than agronomic interest in the early years of the July Monarchy. They were, indeed, of fundamental importance in the lives of France's 33,500,000 inhabitants since they conditioned the domestic output of cereals available for consumption. When related to population numbers they contributed to regional differences in the quality and quantity of human diets and to the pattern of the domestic grain trade, with its striking differences in commodity prices. Isolated pays had to rely almost entirely on the products of their own soil and levels of food consumption were often low as a result.

Contemporaries often wrote of the production and availability of cereals in an all-embracing sense regardless of the different food values of individual crops.[12] Production of eight types of grain has been summated for each arrondissement and amounts required for seed and for consumption in other ways have been subtracted. The most densely populated areas, the poorer environments for grain growing, and those that already specialised in other forms of agricultural activity formed areas of shortage. Seine département, with the major demand centre of 1,106,800 inhabitants, was clearly short of 4,395,000 hl of grain in the average year of the 1830s, and the neighbouring arrondissements of Versailles (−496,000 hl), Melun (−134,000 hl) and Corbeil (−45,000 hl) were also important areas of shortage. Eight other urbanised arrondissements displayed shortages of more than 250,000 hl: Bordeaux (−578,000 hl), Lille (−500,000 hl), Marseilles (−491,000 hl), Lyons (−464,000 hl), Toulon (−365,000 hl), Avignon (−343,000 hl), Rouen (−328,000 hl) and Valence (−281,000 hl). All upland areas were short of grain, as were the densely-populated plains of Flanders, Languedoc, the Rhône and Rhine.

With the exception of arrondissements that contained sizeable cities, all parts of the Paris Basin generated grain surpluses, with the two largest for any part of France coming from the arrondissement of Chartres (+919,000 hl) and Meaux (+584,000 hl). However, the Breton arrondissement of Brest (+580,000 hl) was not far behind. Pastoral areas of Lower Normandy were short of grain but the overall picture for western France, from Brittany through Poitou to the interior of Aquitaine, was one of surplus. Lorraine, Burgundy, Bourbonnais and Berry generated important surpluses but production in some other parts of middle France fell short of demand. In reality, of course, surpluses of one type of grain did not compensate for shortages of another and, although there was some scope for substitution between bread grains, trade was organised and prices were determined by individual crops.

Production of a surplus of any agricultural commodity was pointless unless that surplus could be sold for consumption elsewhere. Major areas of wheat surplus might therefore be expected in three types of location prior to the railway age: in immediate proximity to demand centres (e.g. the core of the Paris Basin); within easy access of internal waterways by land or water (e.g. along the Rhône-Saône system); or with good connections to a seaport for coastal trade or for export (e.g. coastal Brittany and Poitou serving Bordeaux; or eastern Aquitaine being linked to Provence via the Canal du Midi and the Mediterranean). The agricultural population in some surplus-producing areas was itself wheat-eating but in other areas wheat was used as a cash crop and secondary cereals, such as maize, maslin, rye or buckwheat, were consumed by the peasantry. [13]

Nineteen arrondissements produced double or more the volume of wheat that was needed for local consumption (Figure 7.3). These were the famous granaries of France which were normally well served by water routes. Such was certainly the case for Dauphiné, and the Bresse which supplied Lyons and Saint-Etienne; for the export-orientated areas of Finistère; the pays of the Loire valley; and many arrondissements in the inner Paris Basin, although the *beauceron* arrondissement of Chartres lacked a major water link to the capital. Districts in much of the Paris Basin generated large surpluses of wheat, with smaller proportional surpluses coming from western Lorraine, Burgundy, sections of the Garonne basin and much of north-western France. The continuity of this pattern was broken by major cities, stretches of densely populated countryside with important rural industries (e.g. Nord, Lower Maine), poor environments for wheat growing (e.g. the Sologne, Arcoët), and areas such as

the grassland of Lower Normandy that specialised already in
other forms of agricultural production. Because of fundamental
differences in productivity the surpluses from arrondissements
in Aquitaine were much smaller than those derived from the
Paris Basin.

Many arrondissements containing major cities produced less
than half of the wheat consumed by their inhabitants. Such was
most certainly the case for Paris, Lyons, Bordeaux and
Marseilles. In addition, rural areas such as Limousin and the
planèze of Saint-Flour were unable to meet even half of local
demands but the volume of wheat that was lacking was usually
quite small, since secondary cereals figured prominently in local
diets. This was not normally the case for large urban centres.
The département of Seine was reported to be short of 2,950,582
hl of wheat and when the arrondissements of Melun (−180,000
hl), Versailles (−179,000 hl) and Fontainebleau (−1,000 hl)
were considered a total shortage of 3,310,000 hl was recorded
for the core of the Paris Basin (Figure 7.4a). Lyons (−419,000 hl)
and Marseilles (−409,000 hl) experienced almost equal shortages of wheat, as their similarity in population might be taken to
imply. Bordeaux came in fourth position, followed by Valence,

Rouen and Toulon. Sizeable shortages also affected parts of Lower Normandy, Nord, Alsace and the Mediterranean south. Puvis noted that very little grain other than wheat was consumed in the latter region and for that reason Toulouzan remarked 'it is well known that whatever we do we shall not be able to produce sufficient to meet our needs'.[14] By contrast, several arrondissements in the Massif Central were short of only 1,000 hl or so. Chartres arrondissement came in a league of its own among surplus-producing areas, with the granary of Beauce having no less than 505,000 hl of wheat to dispose of after its own needs had been met (Figure 7.4b). Meaux and Compiègne arrondissements followed and no fewer than thirteen arrondissements in the heart of the Paris Basin had over 100,000 hl of surplus wheat apiece. A further score of arrondissements in other parts of France generated similar volumes of surplus, with important clusters in Anjou, Poitou, Upper Languedoc and Dauphiné, and isolated examples being found in Léon, Saintonge, Provence and the Bresse.

The internal flow of commodities was neither controlled nor recorded specially after the Revolution, however contemporary writers did describe the internal wheat trade in the early years of the nineteenth century. Figure 7.5a is based on information from many sources and is impressionistic rather than precise. No attempt has been made to indicate the strength of flows which, in any case, would vary in intensity (and even in direction) from year to year according to the quantity of wheat harvested by region and the possibilities of moving grain between locations. The evidence may be summarised in various ways but there would seem to have been eight components in the wheat trade of the 1830s.

1. Paris formed the major demand centre of the nation and drew upon the wheat resources of at least a dozen départements in the Paris Basin. As Kaplan noted, 'the Paris provisioning machine was a juggernaut encroaching into areas that had once served exclusively the cities of the outer Paris Basin'.[15] The great waterways of the Seine, Marne and Oise, together with the Ourcq canal, played a vital role in channelling wheat downstream to the capital. By contrast, the granary of Beauce relied on road transport and the little river Orge for moving grain and flour to the capital and, in addition, despatched small quantities of grain to Perche and other wheat-deficient pays immediately to the west.[16]

2. The Saône, which was navigable downstream from Gray, and the Rhône connected Lyons and Marseilles with regions such as Burgundy and Dauphiné where important surpluses were produced. Supplies from Burgundy were even sent to Algiers in some years.[17] In bad seasons it was necessary to im-

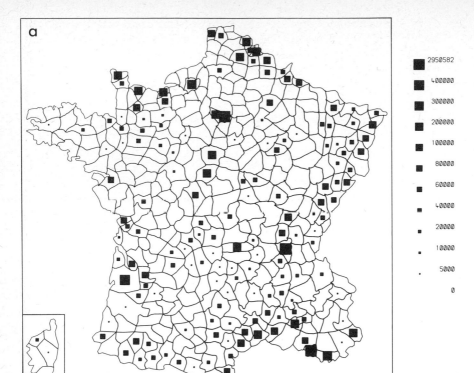

a

■	2950582
■	400000
■	300000
■	200000
■	100000
■	80000
■	60000
■	40000
■	20000
▪	10000
▪	5000
	0

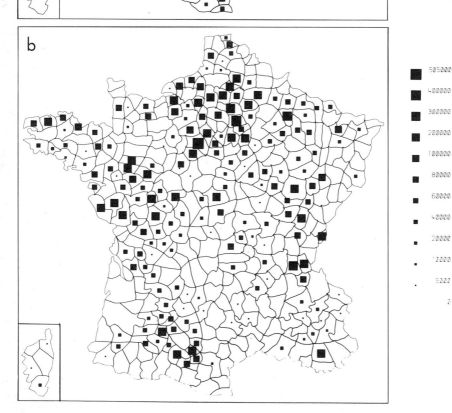

b

■	505000
■	400000
■	300000
■	200000
■	100000
■	80000
■	60000
■	40000
▪	20000
▪	10000
▪	5000
	0

Figure 7.4: Wheat
(a) Areas of Shortage
(b) Areas of Surplus (hl)

port wheat from a variety of Mediterranean sources and these supplies moved northwards along the Rhône in reverse direction to the normal flow. River ports formed important collection points with, for example, Gray on the Saône handling surpluses from as far north as Meurthe as well as from Haute-Saône and Haute-Marne. To the north of Lyons a cluster of river ports performed similar functions for wheat from the Bresse and Burgundy, whilst to the west Beaujeu, Tarare and Thizy marketed wheat from the plain of Forez and even from Bourbonnais before shipment along the Saône and Rhône. These supplies were directed as far south as Ardèche, Gard, Isère, Vaucluse and even Basses-Alpes, being trans-shipped at ports such as Pont-Saint-Esprit and Beaucaire. Dauphiné provided an essential local source of supply for Lyons. Recently constructed canals allowed detailed changes as, for example, in Auxois which could despatch wheat toward Paris or Lyons after 1833 rather than simply supplying local markets. [18]

3. Eastern Aquitaine produced an important wheat surplus which was transported along the Canal du Midi from Toulouse to Agde for sale in the drier Midi méditerranéen and occasionally for export to Spain and Italy. [19] Movement of wheat-laden barges along the canal took five or six days and was always in the one direction in the early nineteenth century, although grain had been despatched westwards down the Garonne from Toulouse to Bordeaux during the ancien régime and, in years when the harvests of Toulouse were slight, wheat had been brought to the Midi méditerranéen from Gascony, Quercy and even from Picardy and Brittany via the port of Bordeaux. *Blés de Toulouse* were shipped from Agde to Marseilles and the ports of Var or else were transported through the Lunel canal to Aigues-Mortes, Saint-Gilles and Beaucaire, thence into départements such as Gard and Basses-Alpes.

4. The Loire valley displayed two distinctive elements in its wheat trade. The lesser of these involved the city of Orléans, which marketed grain from Beauce, Anjou and Poitou for consumption in its immediate surroundings; while Nantes attracted wheat supplies from Anjou, Vendée and even from Bourbonnais and redirected a proportion of these to the Bordelais, to Marseilles and even to England. [20]

5. The Lower Loire system formed part of a wider movement of wheat between ports along the western littoral, with local wheat and supplies from Morbihan, Vendée and the pays charentais being despatched to Gironde, Landes, Basses-Pyrénées and even Portugal. [21]

6. Relatively short-distance movements operated by land or water in many parts of France. For example, wheat was shipped from Côtes-du-Nord to Brest and from the Pays de Caux to

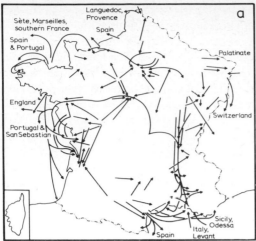
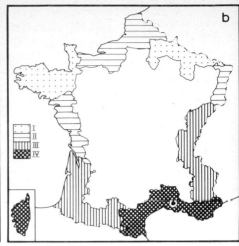

Calvados, and was moved by land from Maine to Perche, from Lorraine to the Vosges, from Picardy to Flanders, from eastern Aquitaine to southern parts of the Massif Central, and from Upper Alsace, Haute-Saône and the Bresse to the Jura.

Figure 7.5: Wheat (a) Internal Trade Routes (b) Price Zones

7. Long-distance movements by sea took surpluses from Brittany to Marseilles, Languedoc, Spain and Portugal, with occasional surpluses from Flanders being directed to similar destinations. [22]

8. Finally, grain was brought to the wheat-hungry districts of Provence and the Lower Rhône from Sicily, mainland Italy and other parts of the western Mediterranean, where they joined the blés de Bourgogne and the more highly-esteemed blés de Toulouse. [23] In similar fashion, wheat supplies from Picardy to Flanders needed to be supplemented by imports brought through Dunkirk.

The configuration of natural routeways together with spatial variations in crop availability provide a basis for understanding the very great differences in wheat prices that were in evidence in the middle 1830s, with values ranging from 10.15F around Thionville to no less than 24.00F around Marseilles and Aix (Figure 7.6). As in earlier times the most marked contrast was between south-eastern and north-eastern France, with outliers of high prices in Lower Normandy, the western Massif Central and Upper Languedoc. Indeed, a spatial classification of wheat prices was incorporated in the legislation of 4 July 1821 which controlled French participation in the international wheat trade. [24] Exportation was prohibited when prices rose above specified département thresholds (ranging from 25 F in Zone I to 19 F in Zone IV). As far as imports were concerned, duties were levied

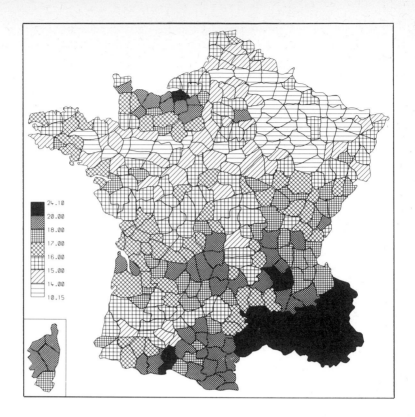

when prices had come down to levels ranging from 26 to 20 F according to zone; and all importation was forbidden when they fell to levels ranging from 24 F in Zone I to 18 F in Zone IV (Figure 7.5b). New legislation was enacted in 1830 but the law of 1821 was re-established in 1832 and was operating at the time of the Statistique.

The economic logic that related wheat availability to prices was simple in general terms but highly complicated in detail. Most parts of Mediterranean France, where cereal yields were generally low, experienced large shortages of wheat which could not be satisfied by southern locations that generated small surpluses. In addition, wheat was the bread grain par excellence in the Midi, with other cereals not being used as a substitute as was the case in some other parts of France. Supplies had to be brought from distant sources and hence prices charged to southern consumers would be raised by the impact of transport costs. This kind of explanation may be applied to other areas of shortage in the middle and upper sections of the Rhône valley and in Franche-Comté. Even where there were important surpluses, for example in the Bresse and Dauphiné, prices were not substantially lower; first, because of the consistent demand for

127

wheat in Lyons, Valence and the towns and countryside of the Midi; and, second, because of the relative ease of communication along the Rhône and Saône and in the Mediterranean Sea itself. In similar fashion, prices in Velay and Upper Languedoc were kept high because these areas could contribute to meeting the demands of Saint-Etienne and Lower Languedoc via the river Loire and the Canal du Midi respectively. High prices in Lower Normandy and Eure reflected the pastoral specialisation of these areas (which displayed large wheat shortages), the fact that parts of Eure practised biennial rotations with important proportions being given over to pulses, that wheat was the dominant bread grain in this densely-populated region with many rural industries, and that some wheat supplies had to be brought by sea (possibly from Brittany) which raised transport costs. [25] The argument to explain high prices in Cantal and Corrèze must be made along somewhat different lines since wheat was not the leading bread grain in these areas, with that role being performed by rye. However, while the peasantry ate black bread the bourgeoisie had a taste for white bread and the necessary wheaten flour had to be brought from the nearest areas of surplus in Aquitaine. Wheat was the leading bread grain in the neighbouring départements of Tarn and Dordogne but movements upslope may have served to keep local prices relatively high in these areas where maize provided a useful supplement and released some wheat for sale.

Conditions of surplus or shortage, differential accessibility to demand centres, regional variations in diet, and the operation of a 'spread-effect' into areas of surplus from neighbouring areas of shortage are the kinds of argument that must be invoked to elucidate variations in commodity prices and they provide plausible enough 'explanations' in many cases. For example, wheat prices in the Bordelais were on the high side (16 to 18 F/hl) largely because this vine-growing pays was an area of wheat shortage which needed to obtain supplies from the interior of Aquitaine, Saintonge, Vendée and Brittany. Bordeaux also performed a tradition role of exporting wheat to the Antilles and this also inflated demand and enhanced prices. Within the basin of Aquitaine there was an interesting contrast in prices to the east and west of Toulouse. Virtually all arrondissements, except Toulouse itself, had wheat surpluses of which to dispose but prices west of the city were 15 F/hl (and occasionally 14 F/hl) whilst those east of Toulouse were 17 or 18 F/hl. The eastern areas could readily despatch surpluses eastwards to Lower Languedoc via the Canal du Midi and the operation of this system inflated prices in the terrefort toulousain but did not extend so strongly into Gers, Lot-et-Garonne or Tarn-et-Garonne from either Lower Languedoc or the Bordelais.

Wheat prices over much of the northern half of France were considerably lower than further south. Lower Normandy formed a major exception that has been discussed already and the harsh environments of the southern Vosges and the Morvan (16 to 17 F/hl) formed minor examples. The arrondissement of Meaux had a surplus of 355,000 hl but it would seem that the strength of demand from Paris raised the local price to 18 F/hl and the same process produced a price of 15 F/hl over much of the Ile-de-France. Prices were lower in arrondissements in Champagne which were not connected easily to the Rhône-Saône trading axis. In addition, these eastern areas lay beyond the radius that was theoretically necessary to satisfy Paris but they were linked to the capital by waterways and had started to despatch grain to Normandy and Picardy after the food shortages of 1817.[26] The Vosges were supplied from western Lorraine and Haute-Marne, and Alsace also depended on these sources and on imports from across the Rhine.[27] Finally, some of the shortages in Flanders and Artois could be met by supplies from Aisne and Somme as well as by local surpluses within Nord and Pas-de-Calais. Prices in these northern demand areas (14 to 16 F/hl) were higher than those in most of Aisne and Somme (13 F/hl).

By far the greater part of France generated surpluses of oats in the 1830s and in some arrondissements of middle France less than half of the quantity produced appears to have been consumed locally. By virtue of the fact that most oats was used as livestock feed it is possible that a greater element of unreality entered these returns than those for grains that were used predominantly for human consumption. Sizeable areas of shortage involved Flanders and Artois, Alsace, Normandy, the Lower Loire, Bordelais, the Pays de l'Adour, the western margins of the Massif Central and the greater part of the Mediterranean south, whilst most of the zone of triennial rotations in northern France generated significant surpluses of this crop. Areas with major concentrations of horses formed the leading zones of shortage, in other words those areas where horse-drawn ploughs were used or large cities or garrison towns were to be found. Seine département (−1,238,000 hl), together with the surrounding arrondissements of Corbeil, Pontoise and Versailles registered a total shortage of 1,876,000 hl which could, however, be satisfied by surpluses from surrounding arrondissements, including Chartres (+295,000 hl), Dreux (+246,000 hl), Compiègne (+234,000 hl), Meaux (+215,000 hl) and Beauvais (+191,000 hl). Shortages in Lille, parts of Normandy and Montmorillon arrondissement (Vienne) could be satisfied from contiguous areas and additional oats were brought to Bordeaux from the pays charentais. Likewise many of the shortages in the

129

Mediterranean south could be made up by surpluses from adjacent areas, with, for example, Arles (+68,000 hl) generating large quantities for sale; but the demands of Avignon, Aix, Marseilles and Toulon could not all be satisfied from the region and must have absorbed part of the large surpluses of Dauphiné. Each arrondissement in Alsace consumed more oats than it produced and took up surpluses from Lorraine and Haute-Saône.[28] Arrondissements in Lower Brittany generated a large surplus of oats which was despatched to Normandy, Flanders and the Midi.[29]

Since oats was not widely used for human consumption the average price was only about half that for wheat, however many of the spatial variations in price noted for the 'noble cereal' also applied to oats. Prices in western Lorraine, Champagne and Picardy were 4.5 F/hl, roughly half those in the south-east (8 to 11 F/hl). Avignon, with a shortage of 139,000 hl, recorded a price of 11.75 F/hl. The cheap northern parts of France, where oats occupied the second season in triennial rotations and where surpluses were produced, lay beyond the spread-effects of Paris, Lille and the deficit areas of Alsace, however the spread-effects of the capital, of Alsace, Franche-Comté and the Bresse raised prices to 6 F/hl in the case of surplus arrondissements of the Ile-de-France and to 7 or 8 F/hl in Haute-Saône, Doubs, Jura and Saône-et-Loire. Oats were expensive in the areas of shortage in Eure and parts of Lower Normandy, with supplies having to be brought in from pays to the south-east, including Beauce. Unlike wheat, a zone of high oats prices was in evidence in the Pays de l'Adour and the central Pyrenees. In fact, prices were quite high throughout the basin of Aquitaine and there were few arrondissements with surpluses of any size downstream from Toulouse.

Rye was produced throughout France but the pattern of areas of surplus and shortage was surprisingly uneven and fragmented. Important though it was in the peasant diet of many regions, rye was not normally eaten by townsfolk nor was it fed in large quantities to livestock. It did not therefore experience the same kinds of demand as either wheat or oats. Contemporary documents made little mention of internal trade in rye and the scale of interaction was limited to local exchanges. For these reasons, arrondissements in Seine département (−112,000 hl) for once did not generate the largest volume of shortage but were surpassed by Saint-Etienne (−126,000 hl) which displayed dietary patterns that were more akin to those of the peasantry of the Massif Central than the townsfolk of the *pays lyonnais*. Shortages were also recorded in the Vosges, Touraine, the *pays blésois* and several parts of the essentially rye-eating Massif Central but most of these could be satisfied by short-distance movements. Lower Poitou and Lower Brittany generated large surpluses, with possibly some being shipped to Bordeaux or the pays charentais or transported along the Loire valley. But regional surpluses were insufficient to meet the needs of Alsace and the Vosges which relied on imports from across the Rhine.

A strong contrast existed between the low price of rye in Champagne, Lorraine and Somme and the high prices in south-eastern France. In Haute-Marne it cost only 6.00 F/hl but around Brignoles (Var) it reached 15.8 F/hl. In addition to this familiar trend it is noteworthy that the Massif Central, the classic rye-eating region, was also part of the zone of high prices with rye costing over 12 F/hl in every arrondissement save those in the Limagnes where surpluses were generated. Many of these upland areas failed to grow all that they consumed and others had only trifling surpluses. In a situation in which substitute cereals were not available locally, transportation across even short distances of highly difficult terrain was very costly. High prices in the pays charentais were similarly a response to local deficits which had to be made good from areas further north.

The pattern of barley shortages and surpluses bore no obvious relationship to the main production areas. Inner parts of the Garonne Basin, many sections of the Paris Basin, Dauphiné and Lower Brittany had disposable surpluses, however the quantities in most arrondissements were small, with Brest (+247,000 hl), Morlaix (+105,000 hl), Arras (+99,000 hl), Chartres (+89,000 hl) and Dijon (+87,000 hl) being exceptions. The northerly distribution of beer manufacture helped explain the shortages in brewing areas along the northern frontier from Hazebrouck to Sedan which together were short of 269,000 hl.[30] Arrondissements immediately to the south were able to satisfy these demands and most internal trade involved short distances, with, for example, large surpluses generated around Dijon being moved into Franche-Comté. Shortages at the mouth of the Rhône were met by surpluses from Dauphiné but the large surpluses of Lower Brittany must have been for export. Prices displayed the predictable regional contrast between the north-east (6 to 7 F/hl) and the south-east (10 to 15 F/hl), with the areas of shortage along the northern frontier and their supply zones in Pas-de-Calais and Aisne having price levels of 8 to 9 F/hl. Shortages in Franche-Comté, Alsace and much of southern France, together with the operation of the spread-effect in supply areas such as Dauphiné and Velay, accounted for high or very high price levels.

Production of maslin appears to have fallen below consumption levels in many arrondissements over the greater part of France. In most instances the quantities of surplus and shortage were small and short-distance flows may have made good deficiencies in southern France. The pattern of prices was very similar to that for wheat, its leading constituent, ranging from 8.00 F/hl around Bar-le-Duc to 22.0 F/hl in Aix and Marseilles, with outliers of high prices in Eure and the arrondissement of Meaux, as was the case for wheat.

Buckwheat was grown in small quantities in many parts of France but large amounts were produced only in Armorica and Limousin where it figured in peasant diets but was fed to poultry elsewhere. Many arrondissements produced virtually what they needed, with only a small margin of latitude. Sections of Brittany generated surpluses which were moved eastward into Upper Brittany and the Armorican marchlands. The unevenness of production, the small quantities entering domestic trade and the dual use of buckwheat as a foodstuff for animals and for humans render differences in price levels very difficult to interpret. Prices were low in the buckwheat-eating region of northern Limousin (5 to 6 F/hl), being higher in Corrèze (8 F/hl) into which supplies had to be brought. In Brittany and its marchlands prices ranged from 11.2 F/hl at Quimperlé to only 5.6 F/hl at Mayenne but showed no obvious relationship between surpluses or shortages.

Maize performed a similar dual role to buckwheat, providing grain for consumption by humans and livestock and green fodder for animals, but it grew successfully in a very different environment. The Pays de l'Adour failed to grow all that it needed and formed the western section of a zone of shortage that included the central Pyrenees and parts of Upper Languedoc. Maize was transported from the terrefort toulousain into the Pyrenean valleys but such supplies were insufficient to satisfy shortages in the Pays de l'Adour. However, Saintonge produced a surplus (+180,000 hl) and part of this may have been shipped to the Adour. This would have created a detached spread-effect and generated high prices in Saintonge (13 F/hl) where maize may have functioned, atypically, as a cash crop. Prices were moderate to high in the central Pyrenees, as befits a zone of shortage, falling to 9-10 F/hl in the Pays de l'Adour and the terrefort toulousain.

Potatoes did not form an important element in agricultural trade, having the disadvantage of considerable bulk and weight in relation to food value, not being able to be stored from one season to the next, and varying greatly in productivity from year to year. There were few large spatial differences between production and consumption, with most arrondissements coming fairly close to an equilibrium situation. In spite of the importance of potatoes in the land-use pattern of north-eastern France, a number of arrondissements in the Vosges and even on the plain of Alsace registered sizeable shortages (e.g.

Weissembourg −202,000 hl, Schelestadt −302,000 hl) but large surpluses were produced in eastern Lorraine and were taken into and across the Vosges. Seine was short of 1,275,000 hl but the inner part of the Paris Basin provided a sufficient surplus to satisfy the capital's needs. Similarly, the shortages in Lyons (−107,000 hl) and Saint-Etienne (−177,000 hl) could be solved by surpluses around Villefranche (+226,000 hl) and Montbrison (+290,000 hl) and in the eastern Massif Central. Such, however, was not the case for Lille (−633,000 hl), Arras (−249,000 hl) or Béthune (−220,000 hl) where shortages could not be satisfied from neighbouring areas of northern France and must have depended on supplies from other parts of the country or, more probably, from Belgium. Large surpluses recorded in Lower Brittany (e.g. +1,132,000 hl around Brest) were despatched through the ports of Brest, Morlaix and Roscoff perhaps to Flanders. [31]

In the Mediterranean south, which was unsuited to potato production, the tubers cost half as much again as the national average (2.10 F/hl) and in some areas reached double that sum. Many southern arrondissements had to be supplied from areas on the margins of the Massif Central and a spread-effect may well have inflated prices to 3.0 F/hl around Alès (+30,000 hl) and Largentière (+100,000 hl) which produced important surpluses. Demands from the Vosges raised prices in eastern Lorraine (1.80 to 2.20 F/hl) and made them substantially higher than in Meuse and its immediate surroundings (1.10 to 1.65 F/hl) which did not have the large market of the north-east quite so close at hand. Potatoes averaged 2.70 F/hl in Paris but the highest prices in northern France were in Beauce and the Flemish interior, where demands could not be met from local supplies. They were cheap in northern and central sections of the Massif Central (falling to 1.00 F/hl in parts of Haute-Vienne) and in Upper Maine with its very large surpluses. High prices in the south-east demonstrated the inadequacy of local supplies and the impact of the spread-effect. Low prices in Meuse and the northern Massif Central reflected situations of surplus, absence of large demand centres, and poor communications which rendered transportation of a bulky and relatively low-value commodity extremely difficult. Surpluses were undoubtedly used as animal fodder, especially for fattening pigs.

As the major sources of grain for consumption by humans and by horses, wheat and oats entered into the circuit of domestic trade during the July Monarchy and, in some cases, were transported considerable distances along land and water routes that served demand centres. By contrast, the five other cereals and potatoes did not enter the commercial circuit to the same extent, being used predominantly to meet local needs. Pays which were dominated by their production, rather than by that of wheat and oats, tended to lie beyond the arteries of trade and the areas of commercial potential that extended from them. Surpluses and shortages were normally small in remoter parts of the country and could be remedied by exchanges between neighbouring settlements and arrondissements. For six of the eight crops there was a sharp contrast between high prices in south-eastern France and low prices in the north-east. Buckwheat and maize were not grown sufficiently widely for such a generalisation to be extended to them. Admittedly, there were important differences in price levels between crops because of differences in use and food value, just as there were detailed variations in the spatial characteristics of high- and low-price areas for the various crops, but the general contrast remained valid none the less. It was a response to differences in productivity, production, consumption, diet, accessibility and the spread-effects that emanated from demand centres. Involvement in commercial farming generated relatively high price-levels in areas of production and of consumption. Such conditions normally did not extend into the less accessible stretches of the French countryside where virtual self-sufficiency was the order of the day. That meant frugality if not downright poverty of diet and the constant threat of disaster should subsistence crops fail.

Notes

1. M. Morineau, 'Y-a-t-il eu une révolution agricole en France au XVIIIe siècle?', RH, vol. 486 (1968), pp. 299-326; M. Morineau, Les Faux-semblants d'un démarrage économique: agriculture et démographie en France au XVIIIe siècle (Paris, 1970).

2. E. Labrousse, 'Les bons prix agricoles du XVIIIe siècle: l'expansion agricole' in F. Braudel and E. Labrousse (eds.), Histoire économique et sociale de la France (Paris, 1970), vol. 2, pp. 367-565; R. Laurent, 'Le secteur agricole' in Braudel and Labrousse, vol. 3, pp. 713-21.

3. Inspecteurs de l'Agriculture, Agriculture française: Nord (Paris, 1843), p. 188.

4. A.A. Monteil, Description du département de l'Aveyron (Paris, 1803), p. 23; J.A.S. Watson and J.A. More, Agriculture, the Science and Practice of Farming, 11th edn. (Edinburgh, 1962), p. 207.

5. D. Faucher, 'L'assolement triennal en France', ER, vol. 1 (1961), pp. 7-17.

6. J.E. Dezeimeris, 'Des moyens d'améliorer l'agriculture en France', JAP, vol. 5 (1841-2), p. 569.

7. M. Nivière, 'Du progrès agricole', JAP, vol. 4 (1840-1), p. 458.

8. A. Young, Travels in 1787, 1788, 1789 (London, 1792), p. 339.

9. M. Planche, 'Vues générales sur l'agriculture', APAPER, vol. 6 (1832), p. 125; Nivière, 'Du progrès agricole', p. 458.

10. A. Meynier, A Travers le Massif Central: Ségala, Levezou, Châtaigneraie, (Aurillac, 1931), p. 110. (Maps relating to aspects of production, productivity and prices for each of the cereal crops are to be found in H.D. Clout, Agriculture in France on the Eve of the Railway Age, PhD thesis, University of London, 1979).

11. G. Jorré, Le Terrefort toulousain et lauragais (Toulouse, 1971), p. 217.

12. J. Peuchet, *Statistique élémentaire de la France* (Paris, 1805).

13. A. Armengaud, *La Population de l'Est aquitain* (Paris, 1961), p. 149.

14. M. Puvis, 'Du climat et de l'agriculture du sud-est de la France', *JAP*, 2nd series, vol. 3 (1845-6), p. 256; M. Toulouzan, 'Considérations sur la direction à donner à l'agriculture de la Basse-Provence', *APAPER*, vol. 6 (1830), p. 145.

15. S.L. Kaplan, *Bread, Politics and Political Economy in the Reign of Louis XV* (2 vols., The Hague, 1976), vol. 1, p. 36.

16. AD Eure-et-Loir. 7 M 15. 'Etat des récoltes en grains'. Letter from prefect dated 16 January 1838; A. Jardin and A.J. Tudesq, *La France des notables, 1815-48* (2 vols., Paris, 1973), vol. 2, p. 172.

17. J. Vidalenc, *Le Peuple des campagnes: la société française de 1815 à 1848* (Paris, 1973), p. 130.

18. F. Lullin de Châteauvieux, *Voyages agronomiques en France* (Paris, 1843), p. 198.

19. R. Forster, *The Nobility of Toulouse in the Eighteenth Century* (Baltimore, 1960), p. 31; G. Jorré, 'Le commerce des grains et la minoterie à Toulouse', *RGPSO*, vol. 4 (1933), pp. 30-72; G. Frêche, 'Etudes statistiques sur le commerce céréalier de la France méridionale au XVIIIe siècle', *RHES*, vol. 49 (1971), pp. 5-43, 180-224.

20. O. Leclerc-Thouin, *L'Agriculture de l'ouest de la France, étudiée plus spécialement dans le département de Maine et Loire* (Paris, 1843), p. 38.

21. J.A. Cavoleau, *Statistique ou description générale du département de la Vendée* (Fontenay-le-Comte, 1844), pp. 710-16; J.P. Quénot, *Statistique du département de la Charente* (Paris, 1818), p. 398.

22. Inspecteurs de l'Agriculture, *Agriculture française: Côtes-du-Nord* (Paris, 1944), p. 162.

23. H. Rivoire, *Statistique du département du Gard*, (2 vols., Nîmes 1842), vol. 2, pp. 252-3; M. Raspail, 'Questions agricoles', *La Flandre Agricole et Manufacturière*, vol. 2 (1835-6), p. 144.

24. Mme Romieu, *Des Paysans et de l'agriculture française* (Paris, 1865), p. 296; Laurent, 'Le secteur agricole', p. 714.

25. L. Musset, 'Observations sur l'ancien assolement biennal du Roumois et du Lieuvin', *AN*, vol. 2 (1952), pp. 143-50; J. Delvert, 'L'évolution économique de la plaine d'Alençon', *AN*, vol. 2 (1952), pp. 263-79.

26. Vidalenc, *le Peuple*, p. 130.

27. P. Leuillot, *L'Alsace au début du XIXe siècle* (3 vols., Paris, 1959), vol. 2, p. 174.

28. M. Marquis, *Mémoire statistique du département de la Meurthe* (Paris, 1803), p. 183.

29. Inspecteurs d'Agriculture, *Côtes-du-Nord*, p. 167.

30. G. Lefebvre, *Les Paysans du Nord pendant la révolution française* (Paris, 1924), p. 193.

31. A. du Châtellier, *Recherches statistiques sur le département du Finistère* (3 vols., Nantes, 1837), vol. 3, p. 20.

Definition

In addition to the arable crops that commonly entered rotations a further dozen 'permanent crops' and 'garden crops' covered 3,476,200 ha. In part these were produced for sale but they also represented essential elements of semi-subsistent polyculture in less commercial parts of the country. The vine was the most widely cultivated special crop during the July Monarchy and displayed at least six important characteristics. As a permanent crop it did not enter rotations. It flourished on sloping and stony ground that was unsuitable for growing cereals; it provided a marketable commodity; and was highly valued. It required intensive labour inputs and supported higher densities of population than would have been possible from cereal cultivation alone. The other special crops possessed some but not all of these characteristics. For example, olives, mulberries, chestnuts and walnuts were tree crops in the same sense as the vine and were thereby excluded from rotations, but hemp, flax and market-garden crops were also excluded, while colza, hops, madder, tobacco and beets were sometimes included in rotations but sometimes excluded.

All but one of the special crops were highly valued, with indexes that divided the proportion of the total agricultural value produced by each crop by the proportion of the agricultural area on which it was grown ranging from 2.08 for olives to 15.77 for mulberries (Table 5.1). Chestnuts formed the exception to this rule, with an index of only 0.33, but as a tree crop they bore obvious affinities with the vine and cannot be omitted from the list of special crops. Neither were chestnuts usually regarded as a cash crop, while all other special crops merited that description. Chestnut trees needed attention but did not require intensive applications of labour however they did help support high densities of population, albeit in their capacity as a subsistence crop rather than by generating outside income. Some special crops were being grown more widely in the 1830s than at the turn of the century (e.g. vines, mulberry bushes, vegetables, colza and beet) but others, such as chestnuts, walnuts, hemp and flax, appear to have occupied less land.

Three fundamental sub-groupings of special crops may be recognised:— 'tree crops' (vines, olives, mulberries, chestnuts, walnuts); high value crops beyond rotations (hemp, flax, market-garden crops); and high value crops sometimes incorporated in rotations (madder, tobacco, hops, colza, beet). The tree crops were predominantly southern in location and often flourished on sites unsuited to cereal cultivation. Most of the remaining special crops were the products of densely populated countrysides. Colza and beet characterised progressive rotations, usually in the northern third of France, but they were also grown in garden plots and for that reason, as well as their high value, have been recognised as special crops.

Taken together, the special crops generated 20.76 per cent of the combined value of crops, pasture and firewood (but not livestock) in the mid-1830s (Table 5.1). Their average index was 3.06, in other words three times greater than might be expected from the 3,476,200 ha they occupied. The index for the vine (2.69) was below the average but the vine occupied pride of place among the special crops, covering no less than 56.7 per cent of the land devoted to them. In addition, 50.0 per cent of the total value of special crops (955,700,000 F) was accounted for by wines (419,029,152 F) plus *eau de vie* (59,059,150 F). Market-garden crops came in second place (157,093,840 F) and represented a further 16.4 per cent of the total value.

Tree Crops

The vine occupied 1,972,332 ha in the mid-1830s but this amounted to only 3.86 per cent of the country. It was rarely grown north of a line running from the Gulf of Morbihan, through Maine, south of Perche and north of Beauvaisis, nor was it widespread enough to merit recording in cool environments in the southern Vosges, Franche-Comté or nine arrondissements of the Massif Central (Figure 8.1). To the south the place of the vine in the land-use balance varied enormously, rising to 31.6 per cent around Béziers and 30.5 per cent around Cognac. In addition to the major *pays viticoles* of Languedoc and Charente, vineyards occupied more than 5 per cent of the surface in Provence, Lyonnais, Burgundy, Alsace and the Limagnes, with quite extensive areas of vines also around Paris and in the Moselle valley.

Nature and culture had interacted across the centuries to produce the quality vineyards of France and much more land was devoted to vines after 1789 since farmers were free to use their land as they wished and past attempts to restrict vinegrowing in favour of cereals had been abolished. Unfortunately the wine produced from these new vines was often poor. This trend was particularly pronounced in southern France where a massive extension of the vine to the detriment of olives and cereals produced a 'veritable revolution' on the plain of the Languedoc after about 1820.[2] East of the Rhône in Provence output was increased by planting vines more densely and viticulture was also being extended in many northern areas.[3] Freedom of cultivation was only one reason for the upsurge of viticulture in the half

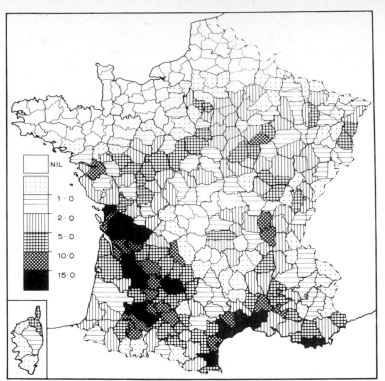

NIL

1·0

2·0

5·0

10·0

15·0

century following 1789. Before the nineteenth century wine had
not usually been consumed in country areas that did not
produce it although it was drunk by townsfolk in most parts of
France. But as time passed and wine prices fell it became an
increasingly popular beverage among the middle and lower
orders of rural and urban society alike and was accepted
gradually as a drink for women and children as well as men. [4] In
areas of intense land fragmentation, such as the plain of Alsace,
vinegrowing enabled high densities of population to be
supported on foodstuffs that could be purchased from elsewhere
with profits derived from selling wine. [5]

The extension of viticulture was criticised widely. In Alsace
the area devoted to the vine had increased by one-third following 1789 with a pronounced reduction in grain growing. As
the mayor of Ribeauvillé remarked in 1823: 'we have too much
to drink but not enough to eat'. [6] Morelot feared that poor *ven-
danges* in Beaune arrondissement would lead to the financial
ruin of the whole area; and in Yonne it was claimed that the rest
of agriculture was suffering because of the extension of the
vine. [7] Land was being taken that should be used for other crops,

labour was absorbed in tending the vine and too great a proportion of scarce manure resources was being devoted to vineyards, leaving insufficient amounts for ploughland. Hence 'poverty remains the normal state in our countryside'.[8] In any case only poor wines were produced, with the inhabitants of Lower Burgundy 'preferring two glasses of bad wine to one of good'.[9] So too it would appear, did many other Frenchmen during the July Monarchy, and the river Yonne provided a convenient route along which mediocre wines were despatched to the capital.

High productivity was far from synonymous with high quality, with more than 40 hl of wine being produced from every ha under vines in many north-eastern arrondissements and around Paris but quality being generally poor so close to the limit of cultivation. In any case, only small proportions of land were devoted to the vine in these marginal areas but vines were planted more densely in northern locations than in the south where summer drought necessitated wider spacing. Densities of 40,000 vines/ha in Champagne and 20,000/ha in Burgundy contrasted with 12,000/ha in Médoc and a mere 4,000/ha in Hérault.[10] Yields were lower further south, with average figures falling below 25 hl/ha in the greater part of the high-quality vineyards of Charentais, Bordelais, Beaujolais and Burgundy. They were equally low in Languedoc and under 15 hl/ha were produced from the rare vineyards of Cantal or Haute-Vienne but such poor yields were not bettered in Périgord or the middle Garonne, where environmental conditions were superior and viticulture had spread on to some of the richest soils of the plain earlier in the century.[11]

Hérault produced the largest volume of wine, with an annual output of 2,615,000 hl of which the greater part came from the arrondissements of Béziers (1,203,000 hl) and Montpellier (933,000 hl). No other arrondissement yielded such large quantities but the very small arrondissement of La Rochelle produced 519,000 hl and the départements of Charente and Gironde produced 1,150,000 hl and 2,017,000 hl respectively. Vineyards in Seine, Seine-et-Marne and Seine-et-Oise yielded as much as 1,124,000 hl, with an even larger amount coming from Aube and Yonne (1,544,000 hl), located south-east of the capital and served by the major waterways of the Paris Basin. According to the Statistique some 930,400 hl of wine (excluding eau de vie) were consumed in Paris arrondissement each year, with 1,407,700 hl being drunk in the whole of Seine. Three arrondissements consumed approximately 400,000 hl apiece each year (Béziers 447,270 hl, Lyons 420,740 hl, Bordeaux 397,820 hl), with an even greater amount being drunk around Blois (518,530 hl).

Different qualities of wine were not distinguished in the Statistique and they can only be treated as a single commodity. With local production subtracted from total supply Seine was short of 1,300,000 hl and the arrondissements of Saint-Etienne (−155,000 hl), Grenoble (−137,000 hl) and Lyons (−83,000 hl) followed at the head of the list. Départements north of the limit of cultivation needed to obtain all the wine that they consumed from elsewhere, for example 254,000 hl in the case of Somme and 183,000 hl for Seine-Inférieure. Unfortunately no information on wine drinking was given for Nord or Pas-de-Calais, which were recorded as beer-consuming areas. The vineyards of Languedoc, Charente and Bordelais generated the largest surpluses, with 3,129,000 hl being available for sale from Gironde, Charente and Charente-Inférieure and 3,079,000 hl from Gard, Hérault, Aude and Pyrénées-Orientales. The vineyards of Var had 1,189,000 hl for sale, with a further 618,000 hl between Beaune and Villefranche being available for trade and the arrondissements of Orléans (+443,000 hl) and Tours (+262,000 hl) also producing large surpluses. Indeed, most areas produced more wine than they consumed and part of the surplus was despatched to major cities, to areas beyond or close to the limit of cultivation, and to uplands in central and southern France.

A national annual surplus of 13,200,000 hl of wine was recorded in the mid-1830s, of which an unknown proportion was used for distilling 1,088,802 hl of eau de vie. In addition, fine wines were transported to long-established overseas markets and also inside France to satisfy the demands of a small and affluent section of society, found especially in Paris. [12] Thus the capital received good wine from the Loire, the pays charentais, Bordelais, the middle Rhône and Côte-d'Or but the greater part of Parisian demands for cheap and moderately-priced wines was met from Seine, Seine-et-Marne and Seine-et-Oise (+244,000 hl) and from départements to the east which were served by the waterways of the Paris Basin. [13] Similarly the needs of the pays lyonnais, Grenoble and Dauphiné were met by surpluses from the arrondissements of Villefranche, Mâcon, Chalons and Beaune which were traversed by the river Saône.

All other areas of shortage within the vinegrowing regions might be satisfied from adjacent supplies but small quantities of fine wine moved greater distances to meet bourgeois demands and some exchange of different types of wine also occurred over short distances. Important quantities of wine from Languedoc were exported by sea and such was certainly the case in Provence where wines from Var were despatched through Toulon to Algeria and for use by the Marine. [14] The wines of Gaillac were better than the average in eastern Aquitaine and were even shipped to the Bordelais. [15] Surpluses from Alsace were in excess of needs in the Vosges and were exported along the Rhine to Switzerland, Germany and the Netherlands. [16] The construction of canals in north-eastern France brought important new features to the trade with wines from the Midi penetrating along the Rhône-Rhine canal and the Burgundy Canal allowing the wines of Côte-d'Or to be shipped more easily to Paris. By contrast with these movements along major waterways, a second order of exchange involved wine being traded between neighbouring pays usually for bread grains. For example, poor and steeply-sloping soils in the Cévennes were relatively useless for growing cereals but were successfully terraced and devoted to 'a kind of monoculture of the vine'. [17] The wine that they produced was exchanged for cereals from Velay and the plain of Forez. [18]

The domestic wine trade was, of course, organised by quality but such information was not given in the Statistique. 'Average' prices by arrondissement are not particularly helpful for such a sensitive and variable commodity, since they reflect not only the intensity of demand but also the unknown quality of the wines involved. Not surprisingly the pattern of average wine prices was quite different from the spatial arrangements of cereal prices. Wine, on average, cost 11.40 F/hl but the price fell to less than 10 F/hl in Upper Languedoc, the middle Garonne, Charentais, Poitou, parts of Haute-Marne and the environs of Nancy and Toul (Figure 8.2). With the exception of the Lower Poitou, each of these areas generated surpluses and each was located beyond easy reach of the capital before the railways were built. At the other extreme, prices exceeded 15 F/hl (and in some cases 18 F/hl) in five types of areas: long-established, high quality vineyards such as Beaune and more especially Bordelais, where overseas demand helped to keep prices high; the capital and its immediate surroundings; areas in Champagne which supplied Paris and where prices were kept high by a 'spread-effect' as well as by high quality; territory close to the northern margin of production, especially areas to the immediate west of Paris that were separated by the capital from eastern supply regions; and finally, a number of southern environments, such as Cantal, that were relatively unsuited to viticulture. The value of production per hectare of vineyard varied enormously, in response to the interaction of variations in productivity, density of planting and wine prices. Values in excess of 300 F/ha characterised not only truly commercial vineyards in Bordelais, the pays lyonnais and Provence but also many parts of north-eastern France where demand, productivity and prices were high but the amount of land under the vine was slight (Figure 8.1).

Unlike the vine, which occupied more land during the July Monarchy than it had done half a century earlier, the surface de-

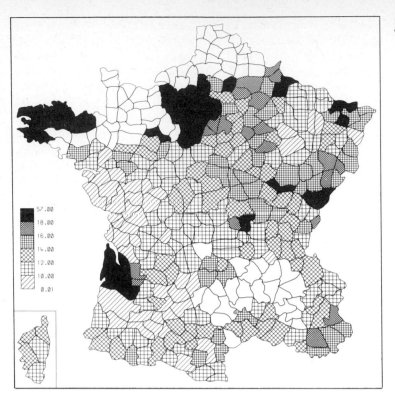

57.00
18.00
16.00
14.00
12.00
10.00
0.01

voted to the olive was on the decline and involved only 121,586
ha in the 1830s in eleven départements from Prades and Carcassonne in the west to Grasse in the east and Valence in the north,
with Corsica as a southern outlier (Figure 8.3). The olive retreated in the early nineteenth century for three main reasons.
First, it had very precise environmental requirements and could
not tolerate frosts. It flourished better on calcareous soils than
on sands, with Toulouzan explaining how north-west-facing
slopes were most appropriate near the coast in Provence,
whereas east- and south-facing slopes were more suitable
inland.[19] Olive trees were most suited to the environment of Var
where they formed veritable forests near Draguignan. They
covered 55,160 ha out of a national total of 121,586 ha and provided 'the pivot of the regional economy'.[20] Olive trees were less
successful further north and to the west of the Rhône where they
were smaller, produced fewer fruit and hence yielded less oil.
Localities open to the south and protected from the north were
most appropriate in Gard, but olive trees responded badly to the
increased evaporation that particularly strong winds
provoked.[21] A series of harsh winters in the early nineteenth
century caused the crop to retreat in Aude, where large numbers

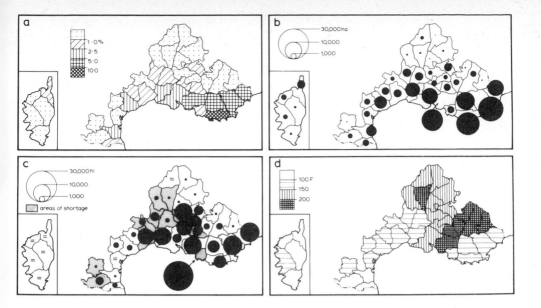

Figure 8.3: Olives
(a) As a Percentage of Land-use
(b) Area under Olives (ha)
(c) Relationship between Production and Consumption (hl)
(d) Price of Olive Oil (F/hl)

of olive trees had been felled prior to the 1840s, and a similar trend occurred in sections of the Rhône valley that were affected by the *mistral*.[22] Competition from other forms of land use provided the second reason for retreat. Thus olive trees were ousted by the vine in parts of Languedoc, while more land on the plains of Provence was devoted to cereals, fodder and dye crops to the detriment of the olive. Third, came cheap imports of olive oil and alternative supplies of oil from relatively new crops such as colza.

Despite the small and declining amount of land involved, the olive formed perhaps the most distinctive crop of the Midi and grew successfully on sloping, parched soils. Total production had an estimated value of 22,776,398 F (or 0.49 per cent of the combined value of crops, pasture and firewood) coming from 0.23 per cent of the land surface. Total production of 167,330 hl fell slightly below the quantity consumed (172,575 hl) each year (Figure 8.3) with the largest net shortage involving the arrondissement of Marseilles (−29,780 hl) which contained important soapworks and other oil-consuming activities. Surpluses from Var and Bouches-du-Rhône could make good that deficit and short-distance movements from the arrondissements of Nîmes and Arles could compensate shortages in Languedoc. However, oil would have to be imported to satisfy demands in Vaucluse (Orange — 10,152 hl, Avignon — 7,889 hl). Prices reached only moderate levels in Vaucluse and the Lower Rhône (150-170 F/hl) but rose to 200 F/hl between Aix and Castellane to the immedi-

143

ate north of the main production areas of Var. Prices were considerably lower west of the Rhône and in Corsica where oil was of poorer quality.

Prior to the Revolution mulberry bushes had been grown in a number of locations, such as Toulouse, Saumur and Grasse, from which they had disappeared by the 1830s when they occupied only 30,438 ha (0.06 per cent of France), but they were often grown on scattered patches of otherwise useless land and the official figure may have been an underestimate. Mulberry growing in Provence and Vaucluse dated from as early as the fifteenth century but production was disrupted between 1789 and 1815.[23] Thereafter a veritable 'fever' of production began in the middle Rhône being stimulated by rising demands for silk among the bourgeoisie of France, Great Britain and the United States. New plantations were made across many types of terrain where silk outworking flourished and where chestnut groves, walnut trees, oaks and even fruit trees were felled to make way for mulberry bushes. All parts of Vaucluse, except the mountains, were growing mulberries and 'the main efforts' of the inhabitants of Ardèche were devoted to their cutivation.[24] There were two foci of cultivation in the mid-1830s, involving the arrondissements of Valence, Tournon and Montélimar to the north and, more especially, a southern concentration around Montpellier, Le Vigan, Uzès (3.07 per cent of land use), Avignon (3.13 per cent) and Alès (5.62 per cent) where mulberry growing was described as more advanced than in any other part of France.[25] The national output of mulberry leaves was valued at 47,779,088 F (0.93 per cent of the combined value of crops, pasture and firewood) from 0.06 per cent of the national land surface and formed the second most highly valued crop.

The same may not be said for chestnut trees which were the second most extensive 'tree crop' in the July Monarchy, covering at least 455,386 ha (0.89 per cent of the land surface) but being valued at only 13,528,190 F (0.29 per cent). Unlike the vine, which provided an important commodity for domestic and international trade, chestnuts were mostly consumed within local, largely subsistent economies, although some high quality nuts were despatched to cities. Chestnuts flourished in relatively impoverished environments with thin, acid, well-drained and often sloping soils, where 'water would not stay on the roots', and which were normally unsuited to other food crops.[26] Hence vast stretches of chestnut trees were described by Malte-Brun in 1833 in Périgord and Haute-Vienne.[27]

Even more important was the diversity of products that the chestnut tree yielded, with Monteil claiming that the twenty varieties of chestnut were the 'most useful trees' to be grown in Aveyron.[28] The fruit of the chestnut provided a source of carbohydrate for man and beast, albeit one with a low calorific value compared with the bread grains which was reflected by its very low value/area index (Table 5.1) Either fresh or dried, chestnuts provided feed for livestock, especially for pigs which could either be killed to enhance essentially vegetarian diets in the countryside, or else be fattened for sale to surrounding areas with bread grains or wine being purchased with the profits. Pigs fattened on chestnuts formed a rare saleable commodity in areas such as Aveyron, with relatively closed economies. [29] In addition, redundant chestnut trees provided an important source of timber for building or for firewood. Rotting leaves and fruits could be used to fertilise soils that produced more valuable crops, such as vines in the southernmost communes of Cantal. [30]

By contrast, the chestnut displayed a number of disadvantages. It needed regular attention and undergrowth had to be cleared systematically if it was to continue to produce effectively. For this reason it did well in isolated environments, such as Rouergue where labour was abundant but once that type of isolation began to break down and labour to become scarce the chestnut groves suffered. [31] Many references cite the existence of a more extensive cover in earlier times but by the mid-nineteenth century interest in chestnut groves had declined. Chestnut trees set in the midst of cultivated fields had the disadvantage of casting shade on surrounding crops but once felled their timber became a source of immediate profit. Such action was condemned as nothing short of vandalism in Périgord, while it was noted that soil erosion on sloping ground in the Cévennes intensified once felling had begun and the value of the remaining tree-covered sections declined. Mulberry bushes replaced some areas of chestnut groves in Vivarais and in Maine-et-Loire further north chestnut trees were felled to make way for cereals. The widespread cultivation of potatoes in mountainous regions assumed part of the subsistence role that had been performed by the chestnut tree and took some of the land it had occupied, however the tree managed to hold its own on very steep 'unclearable' slopes. [32]

There is good reason to believe that the pattern of chestnut groves captured in the Statistique was incomplete. The *arbre-roi* covered only 455,386 ha according to that source (0.89 per cent of the land surface) but the ancien cadastre, which admittedly related to a span of years in the first half of the nineteenth century, registered no less than 563,985 ha. Differences in the rigour of data collection may account for part of the variation between the two sets of figures since, as Arthur Young had remarked, it was difficult to estimate the area covered by dispersed trees. [33] None the less the value recorded in the Statistique for Ardèche (8,139 ha) was only a fraction of that noted in the ancien cadastre (64,032 ha) and Siegfried could write of a 'complete little civilisation' based on chestnut groves between 300 m and 800 or 1,000 m in Vivarais.[34] No matter how imperfect the figures, it is clear that chestnut groves covered extensive areas of Corrèze, Dordogne, Haute-Vienne, Lot, southern Lozère and northern Corsica. Ten arrondissements had more than 10 per cent of their surface under this form of land use, with the most extensive areas recorded around Tulle (42,800 ha) and Périgueux (30,452 ha).

It is unlikely that contemporary statistics on production bore a close resemblance to reality since chestnuts were grown for subsistence rather than for trade. None the less, a range of estimates suggests that annual harvests varied greatly. No less than 3,610,000 hl were estimated to have been harvested in 1815 but in the following disastrous season the figure fell to 841,900 hl. Estimates for 1830, 1832, 1833 and 1835 display great fluctuations (1,431,700 hl, 2,835,500 hl, 2,923,300 hl, 1,848,500 hl respectively), while a yield of 3,478,582 hl was noted in the Statistique, to be counterbalanced by a total consumption of 3,334,091 hl. Most arrondissements, and especially those that were small producers, declared that consumption equalled production. Shortages were recorded in eight arrondissements from Bellac, through Limoges to Brive, in three areas of western Corsica and in two isolated examples in Lozère and Gard but local trade between adjacent arrondissements balanced out these differences.

Chestnuts were used mainly for subsistence but they also served as a cash crop in some areas. For example, wheat was purchased in Lot with the proceeds from chestnut sales and it was declared in Gard that chestnuts formed the only cévénol commodity that might be exchanged for other goods.[35] High quality chestnuts from Vivarais were despatched to Lyons, Marseilles and Paris and *marrons dorés* from Beaupréau and Sègre (Maine-et-Loire) were sold on urban markets but these were exceptions to the general rule. The chestnut remained a subsistence crop to support economies and societies in isolated and predominantly rural areas in central France and Corsica. During the July Monarchy it was on the threshold of decline since these areas were soon to begin to experience exposure to the outside world when new forms of foodstuff would start to become available.

The same conclusion may be reached for walnut trees which covered a mere 6,742 ha in five départements. Dordogne alone contained 5,624 ha of that total, with 2,343 ha in Sarlat arrondissement. These figures probably fell short of reality since Arthur Young had certainly referred to a wider distribution in the 1780s (citing walnut groves in Berry, Quercy, Angoumois,

Poitou, Anjou, Alsace, Bourbonnais and Auvergne) and half a century later walnut trees were still being mentioned in some of these areas. [36] The trees preferred dry, light and slightly calcareous soils and were unsuited to clays and marshy areas. [37] Very hot summers, harsh winters and late spring frosts hampered their development, with foggy, humid springs reducing the hazard of late frost. They needed human intervention to yield satisfactorily and once that intervention was removed their productivity began to decline. Thus the cultivation of walnut trees started to wane in the middle Rhône during the late eighteenth century and their place was taken by mulberry bushes. After the harsh winter of 1789 their numbers declined in Charente, while a fine cover of walnut trees had existed in parts of Puy-de-Dôme in the late eighteenth century but had been removed by the 1840s because harvests had been poor, wood had become more expensive and the scattered trees hindered cultivation in arable fields that surrounded them. Their nuts still provided a useful source of oil for home cooking but by the 1850s walnut trees were reported to be disappearing almost daily. The hardness of walnut wood rendered it particularly valuable for making *sabots*, wine presses, tool handles and furniture, and to satisfy such needs even more walnut trees were felled.

High Value Crops Beyond Rotations

Hemp, flax and market-garden crops shared the common characteristic of being produced intensively on plots that were not normally incorporated in arable rotations. Both textile crops were grown throughout France in the 1830s but both were declining in significance. Hemp growing involved only 176,148 ha to provide a supply of fibre from which fabric was produced usually for domestic needs. The proportion of land involved was small, being only 0.34 per cent of France, with a peak proportion of 2.16 per cent around Angers and an important concentration in the *rieds* of Lower Alsace. To be grown successfully hemp required not only good soil but also heavy fertilisation, hence damper soils in the Ségalas, the Limagnes and Grésivaudan produced better fibre than plots in drier areas, but it had to compete with cereals and vines for very limited supplies of manure. In north-western France hemp was also grown as a cash crop for producing rope and sailcloth. [38] Hemp served at least two other functions, being milled to produce oil for lamps and its residue being used as feed for livestock but new oil crops, such as colza, were performing some of these functions by the 1830s and domestic production of hempen cloth was declining as cotton fabric penetrated the countryside. In addition, the rise of steam navigation was reducing the need for sailcloth. Flax covered 98,241 ha, or roughly half as much ground as hemp, and was absent from many parts of central and southern France. The main concentrations were in Pas-de-Calais (3.69 per cent of Douai arrondissement and 3.47 per cent around Béthune), Nord (Lille 3.55 per cent) and Côtes-du-Nord (Lannion 3.48 per cent), with a smaller area of production in the interior of Aquitaine. By the 1830s the general trend was one of decline.

By comparison with its development later in the century, market gardening was still at an early stage in the 1830s with only 360,696 ha (0.71 per cent of the land surface) being classified as gardens in the Statistique, which were particularly sparse in central France. The financial yield of market gardens was very high, amounting to 157,093,810 F or 3.42 per cent of the value of the combined output of crops, pasture and firewood. By virtue of the range of crops being grown, extensive market gardens were found in close proximity to demand centres, in areas with good access to those centres, and where there were particularly favourable soils or climatic conditions. Thus market gardens occupied between 2 and 3 per cent of twelve arrondissements in the densely populated départements of Nord and Pas-de-Calais, with no less than 9.97 per cent of Seine being market gardens, rising to 14.3 per cent in Paris arrondissement. Minute plots were tended carefully and fertilised heavily with many types of manure, including night soil, to yield a wide range of crops and support 2,000 families around the capital, involving 10,000 people in all. [39] Even so, fresh vegetables were expensive in Paris and Abeausy declared that they were beyond the financial reach of the average Parisian. [40] With every year that passed market gardens were consumed by the fever for new building and the construction of embankments for the railway. Although the circle of communes supplying Paris extended further outwards into Seine-et-Marne and Seine-et-Oise, local supply was unable to keep up with demand since it was reported to take at least seven years to bring a new market garden to productivity. In addition, theft raised serious problems for the gardeners since, 'as everyone knows, the areas surrounding Paris are populated in large part by the outcasts of society'. [41] It was predicted that rising land rents, increasing transport costs, more expensive manure and many other costs would soon ruin the capital's market gardeners.

Fruit and vegetables had been produced for sale in Vaucluse since medieval times, being focused around Avignon, Carpentras and Cavaillon and involving a mixture of cereals, olive trees and other special crops such as madder. Intensive applications of manure were essential in these areas where 'gardening had been pushed to the highest degree of perfection and brought all its advantages'. [42] Roads leading to Aix,

Marseilles, Arles, Avignon and Nîmes were covered with carts, as manure was sought from up to 60 or even 80 km from the market gardens. On his southern travels, Moll gave the splendid illustration of 'a rich cultivator' of Manosque who 'received the dung from his horse in his own hat, placed the dung in a small container on the horse, put the hat back on his head and resumed conversation with his friends'.[43] Women and children, armed with brooms, shovels, baskets and little carts pounced on any droppings that could be found with such vigour that Moll's horse 'was one hundred times the innocent cause of lively quarrels for possession of the precious gift that he had deposited on the road'. Manures and night soil were carted from the towns of the Midi to their surrounding market gardens which were so successful in the environs of Tarascon that 'a single hectare of land sufficed to enrich and to occupy fully a large family'.[44] There were many variations on this theme, with garrison towns such as Metz and Strasbourg being particularly favoured because large quantities of horse manure were produced.

Market gardening also flourished at considerable distances from demand centres provided they were well served by lines of communication. Thus the Val de Loire produced a wide range of vegetables, some of which were sent by river to Nantes and even despatched overseas, and many types of flowers and fruit.[45] In similar fashion Aisne produced vegetables for the capital downstream and the pays of the middle Garonne grew crops for Bordeaux.[46] But the north coast of Brittany provided the most striking example of catering for distant markets. As early as the seventeenth century the inhabitants of Roscoff had cultivated onions on soil so heavily treated with seaweed and marine fertilisers that it was virtually 'artificial'.[47] By the July Monarchy many types of vegetables were grown by between 900 and 1,000 market gardeners (including fifty 'master market gardeners') on plots of roughly 0.5 ha apiece from which three or four crops were harvested each year.

High Value Crops Sometimes Included in Rotations

Five other high-value special crops were included in the Statistique and were usually grown as part of rotations but not invariably so. Fodder beet was not distinguished from sugar beet but both types covered no more than 57,663 ha, a mere 0.11 per cent of national land use. Only in the arrondissements of Valenciennes (6.49 per cent) and Saint-Denis (3.50 per cent) were sizeable areas involved. After an experimental phase during the First Empire sugar beet fell out of favour and continued to be grown only in the Ile-de-France, Upper Normandy, Touraine, Aisne, Yonne, and possibly a few other areas where sugar works continued to operate.[48] In some sections of the Paris Basin sugar beet was only grown out of curiosity in gardens and not until the second half of the nineteenth century was it produced as a field crop. Even less is known about the cultivation of fodder beet during the July Monarchy. Production was reported to be increasing in Lower Normandy, Languedoc and the Flemish coastal plain (where it was ousting flax and beans) but it is impossible to determine how much land was involved.[49] In areas, such as Lorraine, where communal pasture rights remained in operation it could only be grown successfully in gardens and other enclosures and did not enter into regular rotations.[50]

Colza was being grown both in fields and in gardens in the 1830s. Only 0.34 per cent of France (173,506 ha) was devoted to the crop but it was particularly important in Nord, Normandy, the north-east and the Saône valley, occupying as much as 8.9 per cent of the *campagne de Caen* and 10.8 per cent of Arras arrondissement. Towards the end of the eighteenth century colza cultivation had spread from Flanders to Artois, Picardy and Normandy where it entered rotations and occupied part of the land left fallow, and in the early nineteenth century it spread as far south as Lyons. Colza required intensive cultivation and was a good preparatory crop for wheat which gave higher yields after it had been grown. Lullin de Châteauvieux remarked that the quality of oil used in lamps had more than doubled since the Revolution, with candles giving way to oil lamps and a start being made on the lighting of public places.[51] Colza oil satisfied an increasing proportion of that new demand and by the 1830s the crop was starting to be cultivated in regions as far afield as Anjou, Yonne, Vivarais and even Languedoc.

Production of madder was valued at 9,343,349 F, involving no more than 14,674 ha (0.03 per cent of land) of which the greater part was found in Vaucluse, with 5,235 ha in Carpentras arrondissement. A second focus of cultivation involved northern parts of Bas-Rhin. The crop required rich but friable soil which needed to be ploughed deeply. It was grown as a row crop that was propagated by taking suckers from the parent plant. After growing for two or three years the roots were dug up, dried and pulverised to produce the red dye that gave the trousers of French infantrymen their characteristic colour. Madder had formerly been obtained from Izmir and the Netherlands but in 1671 Colbert introduced measures to encourage its cultivation in France.[52] It was accepted initially in Var and was cultivated more widely after legislation in the 1750s exempted the land on which it was grown from taxation. The Armenian Jean Althen settled in Avignon in 1763 and encouraged its cultivation in Vaucluse where it flourished on drained marshlands. Madder seeds were brought from Cyprus and were distributed in

Provence and Alsace with such success that the surplus crop was exported to England. By virtue of the care and attention that madder required, intensive rotations were devised and 'madder became virtually synonymous with "prosperity" in Vaucluse' for almost a century.[53] The crop was adopted in Alsace at approximately the same time and flourished on sandy soils in the Haguenau. Its cultivation was forbidden in favour of cereals during the Revolutionary period and red dyes had to be imported from the Netherlands but in the early nineteenth century it came back in favour.[54] Attempts were made to introduce madder to other intensively cultivated areas such as the Limagnes, however all this prosperity was to come to an abrupt halt when a red chemical dye known as Alizarine was perfected in 1868.

Production of tobacco in France in the 1830s was quite unlike that of any other crop since it was grown only with government authorisation and in six mainland départements, where it covered a mere 7,830 ha, rising to 7,955 ha (0.015 per cent of France) when apparently unauthorised areas in Corsica, growing strong tobacco, are included. Regulations affecting tobacco production had been abolished after the Revolution but were re-established in 1810, 1816 and 1835, whereby tobacco planting was placed under the surveillance of the *régie des contributions indirectes* and was authorised in Nord (572 ha in the 1830s), Pas-de-Calais (441 ha), Bas-Rhin (1,882 ha), Ille-et-Vilaine (555 ha), Lot (1,611 ha) and Lot-et-Garonne (2,769 ha). Production in these areas had been recorded by Arthur Young half a century earlier, when he described tobacco as 'the best crop for producing ready money', and they had included the most extensive areas under tobacco prior to control by the régie.[55] Certainly tobacco yielded as much as 2,000 F/ha in Lille arrondissement but in Lot-et-Garonne the figure fell to 250 F/ha. Part of this variation was due to differences in the quality of the leaf but was also because of differences in the spacing of plants. Such matters, together with the degree of manuring that was permitted, were controlled by the régie.

Even though it was an exhausting crop, tobacco was an excellent preparation for cereals because of the deep ploughing, abundant manuring and careful hoeing that it required. Night soil was used in Flanders as a particularly effective form of fertiliser for tobacco.[56] In the middle Garonne soils which had grown tobacco were used for producing hemp.[57] In northern départements tobacco plants needed shelter from north winds and hence sunny sites in the lee of buildings were best in Flanders and Artois.[58] Farmers in Nord preferred growing tobacco on recently cleared woodland or pasture or on freshly drained marshland with a high humus content. Tobacco produced under such conditions yielded leaves that were both long and broad. By contrast, leaves grown on light soils were less weighty, provided a better aroma and were particularly suited to cigar making. This was the case in Pas-de-Calais and Bas-Rhin, while tobacco from other départements was considered to be better suited to making snuff. Large inputs of labour were required not only for cultivation but also for harvesting at exactly the right moment, drying the leaves and processing them. Not surprisingly, tobacco was a crop of overpopulated pays with highly fragmented land ownership, for example each planter in Lot-et-Garonne had an average of less than 60 ares under tobacco and his neighbour in Lot tended less than 30 ares.[59] In spite of the high returns from tobacco few planters owned special drying sheds and it was normal practice to dry leaves under the eaves of the farmhouse. Tobacco was rotated with other crops in some production areas but heavily fertilised patches could take the crop successfully for a dozen successive years or more.

Hops occupied only 826 ha (0.001 per cent of France) in the 1830s but had the distinction of being the second most valuable crop, coming after mulberries, with the financial yield of a single hectare of hops averaging 1,191 F which was ten times the average value of a hectare of rye, barley, oats or maize. Production exceeded consumption by as much as 531,548 hl in the 1830s and one must envisage sales to Belgium and Germany. Hops were grown in only six northern départements, which formed the major brewing region. The crop grew most successfully in low-lying areas and on gentle slopes that were protected from north and north-east winds and yet received maximum insolation. Once again, freshly cleared land was reputed to give the best results. Very little had been grown in the late eighteenth century, and then only as an object of curiosity, but production increased substantially during the nineteenth century.[60] With the encouragement of the prefect of Haut-Rhin hops were grown more widely on the sandy soils of the Haguenau and Koschersberg after 1810 with plants being brought from Bohemia. Initially German hops were preferred to the home-grown product in Alsace and in north-eastern France hops from Lorraine were preferred to those from Alsace. Not until the 1820s did these prejudices start to be dissipated.[61] A generation later the hop yards of Alsace took a new lease of life and occupied land that had been released by hemp and madder.

Conclusion

The significance of the dozen or so special crops lies beyond the surface of land they occupied since they made relatively little impact on land-use combinations. Rather, their importance stems from their contribution to local diets in relatively 'closed' pays, the income generated from their sale in more 'open' pays and from the employment in country and town that derived from growing, marketing and processing them. Many high densities of rural population would have been impossible without the jobs generated by making wine, distilling eau de vie, brewing beer, making and dyeing linen and silk, and by the many other industrial activities that made use of special crops. On the eve of the railway age some of these were retreating and others advancing, with vines and mulberry bushes ousting chestnut and walnut trees from southern hillslopes and colza and beet advancing at the expense of bare fallows and cereals, albeit to a lesser extent. The course of specialisation had been set even before the first railway line had been built and was to be followed more enthusiastically in the remainder of the nineteenth century as viticulture, market gardening and growing root-crops came to occupy greater roles in the agricultural economy of many parts of France. At the same time, improved internal communications were to bring ruin to chestnut and walnut groves and to flax and hemp gardens as new sources of carbohydrate and oil became available in the countryside. Changes were slow but they were to remove the raison d'être of many rural activities and would in turn contribute to successive waves of depopulation.

Notes

1. R. Dion, *Histoire de la vigne et du vin en France* (Paris, 1959); J. Vidalenc, *Le Peuple des campagnes: la société française 1815-48* (Paris, 1970), p. 236.

2. M. Sorre, 'La plaine du Bas-Languedoc', *AG*, vol. 16 (1907), p. 416.

3. M. Agulhon, *La République au village* (Paris, 1970), p. 29.

4. R. Laurent, *Les Vignerons de la Côte-d'Or au XIXe siècle* (2 vols., Paris, 1957), vol. 1, p. 217.

5. E. Juillard, *La Vie rurale dans la plaine de Basse-Alsace* (Paris, 1953), p. 259.

6. P. Leuillot, *L'Alsace au début du XIXe siècle* (3 vols., Paris, 1959), vol. 2, p. 119.

7. M. Morelot, 'Statistique oenologique de l'arrondissement de Beaune, département de la Côte-d'Or', *AAF*, vol. 29 (1825), p. 229.

8. M. Verrollot d'Ambly, 'Notice sur l'agriculture du département de l'Yonne', *Annuaire Statistique de l'Yonne* (1837), p. 202.

9. Ibid., p. 211.

10. R. Laurent, 'Le secteur agricole' in F. Braudel and E. Labrousse (eds.), *Histoire économique et sociale de la France* (Paris, 1976), vol. 3 (2), p. 676.

11. Inspecteurs de l'Agriculture, *Agriculture française: Tarn* (Paris, 1845), p. 355.

12. Laurent, *Les Vignerons de la Côte-d'Or*, vol. 1, p. 197.

13. AD Seine-et-Marne M 9333. Etat général du mouvement de la navigation dans le département de Seine-et-Marne pendant les années 1838, 1839, 1840, 1841.

14. E. Labrousse (ed.), *Aspects de la crise et de la dépression de l'économie française au milieu du XIXe siècle* (Paris, 1956), p. 327.

15. A. Armengaud, *Les Populations de l'Est aquitain* (Paris, 1961), p. 96.

16. E. Tisserand and L. Lefebvre, *Etude sur l'économie rurale de l'Alsace* (Paris, 1869), p. 157.

17. P. Bozon, *La Vie rurale en Vivarais* (Clermont-Ferrand, 1961), p. 90.

18. A. Siegfried, *Géographie électorale de l'Ardèche sous la troisième république* (Paris, 1949), p. 25.

19. M. Toulouzan, 'Observations générales sur les oliviers de la Basse-Provence', *APAPER*, vol. 1 (1827), p. 371.

20. Agulhon, *La République*, p. 29.

21. H. Rivoire, *Statistique du département du Gard*, (2 vols., Nîmes, 1842), vol. 2, p. 269.

22. Inspecteurs de l'Agriculture, *Agriculture française: Aude* (Paris, 1847), p. 254; P. George, *La Région du Bas-Rhône*, (Paris, 1935), p. 395.

23. D. Faucher, *Géographie agraire: types de cultures* (Paris, 1949), p. 104.

24. AN F. 10 496. Sériculture. Reports by prefects of Vaucluse and Ardèche, dated 1813.

25. E. Robert, 'Industrie séricole', *APAPER*, vol. 12 (1839), p. 282.

26. L.D. de Saint-Christol, *Agriculture méridionale, le Gard et l'Ardèche* (Paris, 1867), p. 43.

27. M. Malte-Brun, *Universal Geography* (Boston, 1833), vol. 8, p. 270.

28. A.A. Monteil, *Description du département de l'Aveyron* (Paris, 1803), p. 38.

29. Anon., 'De l'état de l'agriculture dans le département de l'Aveyron', *AAF*, vol. 17 (1822), p. 144.

30. F. de Parieu, *Essai sur la statistique agricole du département du Cantal*, 2nd edn. (Paris, 1864), p. 76.

31. R. Béteille, *La Vie quotidienne en Rouergue au XIXe siècle* (Paris, 1973), p. 38.

32. J.R. Pitte, 'Les origines et l'évolution de la châtaigneraie vivaraise', *Comité des Travaux Historiques et Scientifiques, Bulletin de la Section de Géographie*, vol. 82 (1978), pp. 165-78.

33. A. Young, *Travels during the Years 1787, 1788 and 1789* (London, 1794), vol. 2, p. 62.

34. Siegfried, *Géographie électorale Ardèche*, p. 26.

35. B. Mazières, 'Etude géographique de l'alimentation dans le Lot entre 1840 et 1880', *RGPSO*, vol. 25 (1953), pp. 293-312; Rivoire, *Statistique du Gard*, vol. 2, p. 231.

36. Young, *Travels*, vol. 2, pp. 61-2.

37. E. Huard Du Plessis, *Le Noyer: traité de sa culture*, 2nd edn. (Paris, 1867), p. 15.

38. M. Chasle de la Touche, 'Essai sur la culture du chanvre dans les départements de l'ouest de la France', *AAF*, vol. 32 (1825), p. 201.

39. M. Phlipponneau, *La Vie rurale de la banlieue parisienne* (Paris, 1956), p. 74; O. Leclerc-Thouin, *Compte-rendu des travaux de la Société Royale et Centrale d'Agriculture* (Paris, 1844), p. 16.

40. M. Abeausy, 'De l'état de la culture maraîchère aux environs de Paris', *JAP*, vol. 3 (1839-40), p. 452.

41. Ibid.

42. L. Moll, 'Engrais-Jauffret', *Bulletin de la Société Libre d'Agriculture du Gard*, vol. 7 (1837-9), p. 23.

43. Ibid., pp. 24-25.

44. H. de Villeneuve and E. Robert, 'Revue agricole de la Provence', *APAPER*, vol. 12 (1839), p. 54.

45. M. Millet, 'Descriptions des fleurs et des fruits nés dans le département de Maine-et-Loire', *Mémoires de la Société d'Agriculture, Science et Arts d'Angers*, vol. 3 (1835), pp. 5-220.

46. Anon., 'Statistique description de l'Aisne', *AS*, vol. 2 (1802), p. 179; P. Deffontaines, *Les Hommes et leurs travaux dans les pays de la moyenne Garonne* (Lille, 1932), p. 185.

47. H. de Thury, 'Cultures maraîchères de Roscoff', *AOF*, vol. 4 (1845), p. 498.

48. H.D. Clout and A.D.M. Phillips, 'Sugar beet production in the Nord département of France during the nineteenth century', *Erdkunde*, vol. 27 (1973), pp. 105-19.

49. P. Farel, 'Notes sur la culture de la betterave dans le midi de la France',

APAPER, vol. 6 (1832), pp. 225-29; R. Blanchard, *La Flandre*, (Dunkirk, 1906), p. 299.

50. M. Commerrell, 'L'état de l'agriculture du canton de Puttelange', *Mémoires de la Société Libre d'Agriculture de la Moselle*, vol. 1 (1803), p. 36.

51. F. Lullin de Châteauvieux, *Voyages agronomiques en France* (Paris, 1843), p. 306.

52. M. Gerrier, 'Discours', *Nouveaux Mémoires de la Société des Sciences, Agriculture et Arts du Département du Bas-Rhône*, vol. 2 (1834-5), p. 238.

53. A. Peeters, 'Les plantes tinctoriales dans l'économie du Vaucluse au XIXe siècle', *ER*, vol. 60 (1975), p. 54.

54. D. Succ, *Essai sur la garance*, (Paris, 1861), p. 6.

55. Young, *Travels*, vol. 2, p. 80; J.A. Barral, 'Culture et monopole du tabac', *JAP*, vol. 1 (1843-4), pp. 299-304.

56. E. Marie, 'Agriculture du département du Nord', *JAP*, 2nd series, vol. 3 (1845-6), p. 101.

57. P. Deffontaines, *Les Hommes et leurs travaux*, p. 210.

58. J. Cordier, *Mémoire sur l'agriculture de la Flandre française et sur l'économie rurale* (Paris, 1823), p. 330.

59. G. Heuzé, *La France agricole: région du sud-ouest* (Paris, 1868), p. 99.

60. R. Zeyl, 'La culture du houblon en Alsace', *AG*, vol. 39 (1930), pp. 569-78.

61. J. Chiffre, 'Evolution et transformations d'un type d'agriculture familiale: la culture du houblon en Côte-d'Or', *RGE*, vol. 14 (1974), pp. 393-410.

The Legacy

France remained firmly in the timber age during the July Monarchy even though the extraction of coal increased from about 1,000,000 tonnes on the eve of the Revolution to 2,500,000 tonnes in the mid-1830s and canal construction allowed it to be moved more easily and cheaply. [1] For example, the canal network of Nord was enhanced considerably between 1820 and 1830 and by 1833 the Oise was canalised which facilitated shipments to Paris. [2] In spite of such changes in transportation and fuel supplies France's iron- and glass-works continued to display many old-established locational features, with charcoal retaining its importance in iron-making districts such as Moselle, Haute-Marne and Ariège. Timber was also of vital significance for fuelling a host of other industries, for shipbuilding, constructing dwellings, heating homes and providing raw material for craft activities. As well as providing timber for felling or gathering many woodlands also yielded pastoral resources such as beech mast, acorns and the valuable light mantle of ground vegetation that covered relatively open areas. In addition, herbs and wild fruits were collected by the poor who also cut undergrowth for kindling and for fertilising their plots.

As Sir John Clapham remarked, 'the most universal and most essential type of common was the common woodland' and for that reason communal rights of woodland ownership and exploitation were not abolished at the Revolution. [3] They remained in operation during the July Monarchy but were abused extensively, as in Var where timber was felled indiscriminately, goats were let loose to graze, and fires were started deliberately by shepherds who wished to burn trees and shrubs so that flocks might graze the fresh new vegetation that followed. [4] Many fires became uncontrollable and ravaged what remained of the département's woodlands, especially those in communal ownership. Peasants even erected temporary cabins as they felled the few examples of high timber that were still to be found.

Clearance before, during and immediately after the Revolution brought formidable reductions to the areas under trees. Feudal woodlands, wooded country parks and even isolated trees suffered during the Revolutionary fury, whilst sale of *biens nationaux* contributed to further deforestation for agricultural purposes. This trend continued vigorously into the nineteenth century in response to rising population numbers and the desperate need to grow more food. Deforestation of fragile environments was widespread and was roundly condemned by agricultural writers, as in Drôme where slopes were cleared that 'were destined by nature' to be wooded, and in Aveyron where défrichement on hillslopes was 'a kind of barbarism very much worse than that of savages, who do not cut down trees and destroy the soil that nourished them'.[5] Superb woodland in Limousin had given way to 'thin pastures grazed by miserable flocks'.[6] Mature trees were felled indiscriminately after 1789 in many areas, such as Charente, where they were in great demand for shipbuilding, and the quality of that département's remaining timber resources declined severely.[7] In Velay the felling of whole forests of fir trees was stimulated by high grain prices in 1817-8 and the need to produce more food but also by the insatiable demand for planks, beams, pit timber and firewood emanating from Saint-Etienne and Lyons. Around Yssingeaux it appeared that 'we are destroying the future of our forests for Saint-Etienne'.[8] Clearance was not just for extending the agricultural surface and obtaining wood but also for enlarging areas of rough pasture. It was estimated that four times as much woodland had been cleared for that purpose as for cropping; and the trend was profoundly abhorred.

No less than 8,804,550 ha, one-sixth of the land of France, was declared to be tree-covered in the mid-1830s but no record of hedgerow or windbreak timber was made, even though this provided a useful source of wood in the *pays bocagers* of western and central France (Figure 9.1). The tree cover was uneven in the extreme, with less than 10 per cent and often less than 5 per cent of the land in arrondissements in Armorica being wooded (Figure 5.2a). Such regions contained large stretches of moorland that were grazed by 'wretched flocks of sheep . . . Planted with trees, these areas would double the fortunes of their owners'.[9] By contrast, more than a quarter of Alsace, the Vosges, parts of Lorraine, Burgundy, Franche-Comté and Provence were wooded. Sixteen arrondissements, all but Toulon and Brignoles being in the north-east, had more than 35 per cent of their land under trees, with a maximum of 51.6 per cent being around Saint-Dié. Important sections of the Landes were covered with immense pine forests that were so isolated that 'the presence of man is virtually unknown'.[10] It is not coincidental that the major remaining reserves of timber during the July Monarchy were at considerable distances from shipbuilding ports although they were coming under pressure from shipbuilders who sought timbers from accessible parts of north-eastern départements such as Moselle and eagerly awaited the possibility that railways would further open up their resources.[11]

One-eighth of the country's woodland was owned by the state

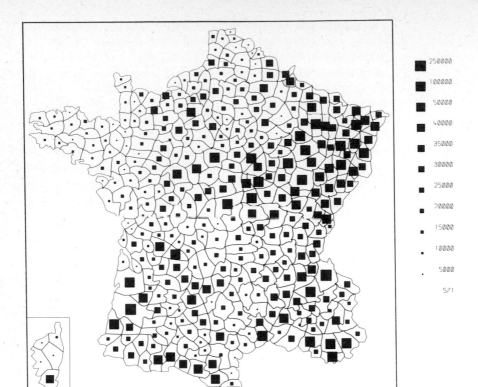

■	250000
■	100000
■	50000
■	40000
■	35000
■	30000
▪	25000
▪	20000
▪	15000
·	10000
·	5000
	571

or formed part of the royal domain (Table 5.1) but the latter occupied only 52,972 ha, entirely in the Ile-de-France. The remaining seven-eighths were owned by private individuals or were held collectively by communes or sections.

Access to timber was a highly important right for members of any rural community and special legislative measures were taken in 1827 to ensure the survival of common woodland while other forms of commonland were gradually being alienated. This did not prevent the quality of common woodlands declining markedly. Unfortunately communal woodlands were not distinguished from private timber in the Statistique but an enquiry by Herbin de Halle in 1834 estimated that 1,802,482 ha were owned by 11,448 communes and a later source, which excluded Alsace and Lorraine, mentioned 8,157 communes and 2,877 sections as owning woodland.[12] The remaining 368,705 ha of 'woodland' were described as sols forestiers, which were found in twenty-four départements of which Hautes-Alpes (100,000 ha), Nièvre (52,616 ha), Landes (39,721 ha), Ariège (23,238), Basses-Pyrénées (20,752 ha) and Hautes-Pyrénées (17,756 ha) accounted for a very large part. Areas of sol forestier were treeless stretches within forest perimeters from which no firewood production was recorded. No details of sols forestiers were given at the arrondissement scale and they, together with 32,978 ha of *makis* recorded in Corsica, have therefore been allotted to the 'other' land-use category defined in Chapter 5.

State forests were not found throughout France and were indeed absent from large stretches of Armorica, the Garonne basin and the Alps and from parts of Picardy and the Massif Central. By contrast, more than half the woodland was state property in a block of arrondissements in the highly wooded north-east and in four other arrondissements elsewhere in the country, namely Montluçon (Tronçais forest), Valenciennes (Saint-Amand forest), Yvetot and Mamers. State woodlands were the responsibility of a director in Paris and thirty-two regional conservators who employed a workforce of 9,400 to ensure a rational programme of thinning, felling and replanting in harmony with environmental conditions, commercial requirements and local custom.

Yields of firewood varied from forest to forest in accordance with the type and age of the trees, with local climatic and pedological conditions (which affected growth rates), and with the degree of wise management or of devastation that had occurred in the past. None of these features was recorded but the Statistique indicated that the national average yield from communal and private woodlands was substantially below the national average for all types of woodland. The condition of communal and private woodlands varied enormously, with long established communal *usages* having suffered serious abuse in many parts of the country. It was claimed that France had passed not only through an initial phase of uncontrolled exploitation but also a second phase of careful management following Colbert's forest ordinance of 1669 which established a legal and theoretical basis for grazing, thinning and clearance.[13] The woodlands suffered serious damage in the Revolutionary years but it was argued that a third phase of 'forest farming' had been reached soon after 1800, whereby attention was paid to pruning, tree selection, land drainage and building of forest roads. Some parts of France may have entered that phase, but the record of contemporary observers during the July Monarchy shows it to have been far from the case in most départements. A fourth phase had already started in Germany and was characterised by the afforestation of moorland which French foresters sought to emulate.

Deforestation was not recorded with any precision during the early nineteenth century. A total of 47,030 ha were authorised for clearance between 1828 and 1836, with fourteen départements close to demand centres or navigable waterways each having more than 1,000 ha authorised.[14] With the exception of Var, each was in the northern, more industrialised half of France, with authorisations in Saône-et-Loire, Meurthe, Somme and Oise surpassing 2,400 ha apiece. An estimated 109,049 ha received permission for clearance over the longer span between 1803 and 1834, with an average of 3,500 ha disappearing each year. By the early 1830s the rate had accelerated with about 5,500 being authorised annually but, of course, it is not known how much private or communal woodland was cleared without permission. Most writers were extremely critical of the quality of the remaining French woodlands in the 1830s. The third phase may have been entered in state woodlands in some parts of the country but the legacy of past mismanagement was only too apparent over the greater part of France.

Local details varied but the general message was that extensive devastation had occurred in the previous half century. Fragile upland environments, such as those of Basses-Alpes, suffered most. In 1831 Baudrillart reported that almost all of that département's woodlands had been removed and that this had accelerated soil erosion from hillslopes and increased rates of deposition in the valleys.[15] Small flocks could enter communal woodlands unnoticed and were even tolerated by some *gardes champêtres*. Transhumant livestock provoked more destruction and the shortage of firewood meant that the peasantry took whatever communal timber they could find.[16] They even burned animal dung and thereby reduced the meagre amount of manure available for fertilising the soil. In the coldest winters they lived alongside their animals to share their body warmth. So too did mountain folk in parts of Puy-de-Dôme where firing was so scarce that bread was baked only a couple of times during the winter months and even then heather and animal dung were used to heat the ovens.[17] Much of the beauty of Upper Provence had gone because of soil erosion and the increased aridity which was alleged to have followed deforestation.[18] Similarly in Aude the amount of rainfall and the number and abundance of springs was claimed to have diminished after widespread clearance of woodland and the *garrigue*.[19] Reports of flooding following défrichement were also widespread and came from environments as diverse as Roussillon and Franche-Comté. As early as 1820 government purchase of the most eroded uplands was being advocated.[20] Replanting was needed urgently around large towns like Marseilles where woodlands had been thoroughly ravaged for firewood. Communal woodlands were usually devastated to a greater degree than state or private forests so that in Anjou, for example, high timber was no longer found on commonland.[21]

Productivity and Prices

The poor condition of much communal and private woodland was reflected by the fact that, on average, only 3.99 m³/ha of firewood were produced, by comparison with 4.96 m³/ha from state forests and 5.55 m³/ha from royal forests where management codes were enforced more strictly. In only a score of arrondissements did the yields of communal and private woodland exceed 8.0 m³/ha, with very high yields recorded around Mamers (16.72 m³) and Saint-Quentin (13.22 m³) but average yields were less than 1.5 m³ across a great stretch of southern France, where environmental conditions were far from ideal for tree growth, and in Seine-Inférieure, where woodlands were under great pressure from domestic and industrial consumers (Figure 9.2a). State forests in thirty-one predominantly northern arrondissements yielded more than 8.0 m³/ha (Figure 9.2b). By virtue of their high productivity state forests around Bar-le-Duc (11.92 m³) were stated to generate no less than 153,795 m³ of firewood each year from 12,892 ha, being surpassed in volume by only Sarrebourg (179,970 m³ from 32,954 ha, with an average yield of only 5.46 m³) and Saint-Dié (209,835 m³ from 44,749 ha, with an average yield of 4.68 m³). At the other extreme, four Pyrenean arrondissements (Bagnères, Saint-Gaudens, Foix and Saint-Girons) contained more than 13,000 ha of state forest apiece but yielded only 1.24 m³/ha on average. Output was even lower in the ravaged woodlands of Var, where each arrondissement produced less than 1.0 m³/ha and around Toulon the figure fell to 0.22 m³/ha. Puvis described the département's woodlands as nothing better than 'steppes', only one-tenth of which produced wood, with the remainder being abandoned to scrub that was devoured by flocks of sheep and goats; while in neighbouring Languedoc there were no longer any 'forests worthy of the name'. [22]

The consumption of firewood in many arrondissements was not shown in the Statistique and, in any case, figures were only given for private and communal woodlands and related to consumption by *usines*. Nothing was noted for state woodlands or for domestic consumption. Hence it is not possible to depict production and consumption and suggest the likely nature of internal trade in firewood with the same degree of precision that was possible for cereals. Industrial consumption of wood from private and communal woodlands was, however, noted for each arrondissement in twenty départements. Only five départements recorded shortages and, with the exception of Seine (−1,200,000 m³), these were very small in volume. The pattern of firewood prices reflected the delicate interaction between forces of supply and demand, with supply being conditioned by

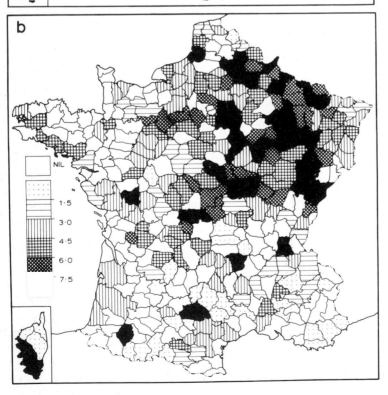

Figure 9.2: Productivity
(a) Private and
Communal Woodland
(b) State Woodland
(m³/ha)

a

1·5
3·0
4·5
6·0
7·5

b

NIL
1·5
3·0
4·5
6·0
7·5

the extent and quality of woodland and the availability of float-able waterways, and demand being a manifestation of the degree of urbanisation and industrial development and also being conditioned by climatic differences produced by latitude and altitude. On average, firewood cost 6 F/m^3, with a range from less than 4 F in parts of the Massif Central, the interior of Brittany and parts of the south-west, rising to 16 F in the urbanised arrondissement of Paris and in the environs of Saint-Denis and Narbonne (Figure 9.3). In four areas it exceeded 12.0 F, namely Paris (16.0 F) plus four neighbouring arrondissements; Marseilles (12.9 F) and Arles; Dunkirk (15.0 F) and Hazebrouck; and a stretch of territory from Narbonne (16.0 F) to Montauban. Only in four other widely dispersed districts did prices reach 12.0 F (Mâcon, Arcis-sur-Aube, Bayeux and Retel).

Prices were obviously highest in and around the capital, where the volume of domestic and industrial demand was greatest, and in cool northern areas such as Flanders, Picardy, Upper Normandy and the Ardennes, with important cities and industrial activities. Climatic conditions were more benevolent in the Midi but firewood remained virtually the sole source of fuel for southern industry and was of great importance for

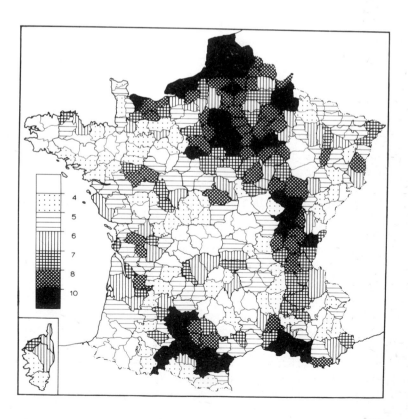

Figure 9.3: Average Price of Firewood (F/m^3)

domestic heating in cool periods of the year. Mediterranean woodland resources were poor and hence firewood prices were high, especially at the mouth of the Rhône and in Upper Languedoc. Firewood was also expensive throughout the predominantly arable pays of the Paris Basin where the 'spread-effect' of Parisian demands made itself felt as it did in eastern, more wooded areas that were linked to the capital by waterway. Prices were high along the valleys of the Saône and Rhône not only because of demand from the cities and industries that they contained but also because they channelled wood from more accessible sections of north-eastern France to firewood-hungry pays further south. Prices were average or below average throughout much of the well wooded north-east, especially in sections of Franche-Comté, Haute-Marne and the Vosges that were beyond easy access by waterways leading to Paris or to the south. Price levels were normally well below the national average in the predominantly agricultural western half of France since the volume of demand was less, as was the degree of urbanisation and the physical possibility of floating firewood over long distances. With few exceptions, the needs for firewood in arrondissements of central and western France must have been met from the immediate locality. The availability of coal as an alternative fuel in the pays lyonnais and inner parts of Nord may help account for the relatively low firewood prices around Lyons and Lille. More substantial timber was certainly still in great demand in Nord during the July Monarchy and Belgian shipbuilders purchased stout timbers from the département's forests.

Waterways were of crucial importance for transporting firewood and the patterns of prices and values of production are only intelligible in relation to that fact and the operation of 'spread-effects' upstream from demand centres. The headwaters of the Seine and Marne carried wood from Nivernais and eastern sections of the Paris Basin to the capital in the form of rafts or boats that could not be returned upstream but were broken up and burned as firewood in Paris. [23] Timber from Upper Alsace was floated down the Rhine from Colmar to the Netherlands and firewood from the Côte-d'Or was despatched both to Paris and along the Saône towards the Midi. [24] Further south, timber from woodlands in the eastern Massif Central was used locally in the mines of Firminy, was carted eastwards to Saint-Etienne and was also floated down the Loire to Orléans and Nantes, to be joined en route by wood from Cher and other surrounding pays. [25]

Canal construction was bringing changes in detail to the movement of firewood to demand centres. For example, the Burgundy Canal opened the woodlands of Auxois and Armançon to the Paris market. Wood from these areas had previously been used for making charcoal for local industries which could now be supplied with coal brought by canal from Givors and Epinal. Marseilles was supplied from surrounding parts of Bouches-du-Rhône and Var and from Corsica, but the lack of security on the island and poor communications with the mainland rendered timber exploitation such an inefficient operation that 'Swedish wood arrived in Toulon more cheaply than wood from Corsica'.[26] In western France the river Charente served for moving wood to the naval port of Rochefort.

The 'spread-effects' upstream from urban demand centres are also reflected in the total value/ha of the firewood that was produced since high prices or high productivity, either alone or in combination, generated high values of production. Since state woodland displayed higher average productivity its firewood also had a higher mean value/ha (31.25 F) by comparison with communal and private woodland (23.55 F). Crown woodland in Seine-et-Marne averaged 52.75 F/ha but values for the other three départements of the Ile-de-France were not given. State woodlands yielded firewood worth more than 60 F/ha in twenty-six arrondissements with peak values being in the northern arrondissements of Saint-Quentin (189.75), Abbeville (106.70), Vouziers (104.50) and Sedan (105.15). Yields valued at more than 40 F/ha involved many arrondissements throughout the northern and eastern sections of the Paris Basin. State forests in the Ile-de-France were ideally located to supply the capital and those in Picardy and the Ardennes could meet the needs of their densely-populated environs and also despatch firewood to the capital. Values were also very high in areas east of Paris that were linked to it by river, in the pays of the Saône and in Vienne arrondissement adjacent to Lyons. They were not greatly above average in much of north-eastern France which was not accessible to Paris or to the Saône valley.

The absence of state woodland from so many arrondissements creates a necessarily fragmented picture but a much more comprehensive view is to be derived from communal and private woodland. Only nineteen arrondissements generated firewood worth more than 60 F/ha and peak values involved Saint-Quentin (117.90), Coulommiers (99.90), Vouziers (95.15) and Sedan (88.65). Values in excess of 30 F/ha were encountered in most areas north of a line between Côtes-du-Nord and Dauphiné. These areas contained major urban centres and industrial activities. Only rarely were values of more than 30 F/ha found in the south and these were in close proximity to Bordeaux, Lyons and Toulouse. As in the case of state forests, and for the same reasons, communal and private woodlands in the Ile-de-France, Picardy and the Ardennes stood out with particularly high values of more than 50 F/ha. These areas were joined by the woodlands of Anjou, upstream of the shipyards and industries of Nantes. Relatively low values involved the Vosges but especially the densely populated regions of Flanders and Upper Normandy, where productivity was exceptionally low (less than 1.5 m³/ha) even though industrial and urban demands rendered prices high. Imperfect though the record of consumption may be, evidence on firewood prices and the value of firewood yields provides further substance for distinguishing northern France, with relatively well-developed commercial circuits serving Paris and to a lesser extent Flanders and the Lyonnais, from the 'other France' where commodity movements were less pronounced. Cities such as Nantes, Bordeaux, Toulouse and Marseilles did, of course, extend their demands into their surrounding countrysides but their supply areas were in a completely different league from the vast area that came under the commercial influence of the capital.

The Third and Fourth Phases

During the July Monarchy most French woodlands bore the scars produced by devastation and mismanagement, however some progress was being made in restoring the legacy of the past and also increasing the amount of land under trees. For example, in the Vosges there were fine examples of pines being planted on steep and rocky terrain and voids had been repopulated where extensive areas of woodland had been felled to fuel saltworks. [27] Sections of the Landes had already been planted with pine trees and wide areas were being actively reafforested in the 1830s. The département of Landes was reported to contain only 127,000 ha of woodland in 1821 but by 1832 the surface had increased to 215,000 ha and 225,000 ha were recorded in the Statistique, with a further 39,720 ha of sol forestier. [28] The national picture of afforestation is not known in detail but by 1847 Martinel could report than an important share of the heaths of Champagne had been afforested successfully. [29] This was a very important improvement since many experiments in the late eighteenth and early nineteenth centuries had resulted in ruin. Landowners in the Sologne 'had started planting forests in a pays where cereal cultivation was ruinous' and, in de Chambray's words, 'a revolution had started and was really making itself felt' as worn-out ploughlands and stretches of heath were planted with pines. [30] In Brittany, the Paris-based Compagnie de Bretagne attempted to plant up sections of moorland in 1829 and 1830 but encountered formidable opposition from the local peasantry. [31] Lullin de Châteauvieux estimated that it would be advantageous to plant up no less than 1,500,000 ha throughout the country. [32]

French foresters looked enviously at silvicultural practices in

Germany and expressed their regret that such a modest start had been made in planting new forests in France and improving what remained of the old. The inhabitants of upland France displayed strong opposition to afforestation schemes as they feared that their rough grazing land, poor though it was, might disappear beneath a mantle of trees. Attempts by officers of the *Eaux-et-Forêts* to apply forestry codes more effectively contributed to the so-called *guerre des demoiselles* in Ariège in 1829 which was followed by outbreaks elsewhere in the Pyrenees, Alps and Jura. [33] In such circumstances destruction of fertile valley floors by stones that were swept down from eroded mountain slopes continued as before and districts 'so invaded wear the appearance of desolation and death'. [34] As Blanqui was to show in his report to the *Institut de France* in 1846 such areas had become virtually worthless but could be given value once again if they were planted with trees. [35] Already it was as if 'Barcelonette, Embrun, Verdon and Devoluy are a sort of Arabia Petraea in the Hautes-Alpes'. [36] Unless afforestation were started under emergency legislation 'in fifty years there will be such another desert between France and Piedmont as there is between Syria and Egypt'. In fact a 'veritable crusade against deforestation' did begin at mid-century and regulations for woodland management were imposed more strictly first in state forests and later in areas of communal woodland but only with great difficulty since local opposition to afforestation remained strong.

Notes

1. A. Jardin and A.J. Tudesq, *La France des notables, 1815-48* (2 vols., Paris, 1972), vol. 1, p. 206.

2. M. Chevalier, *Des Intérêts matériels en France* (Paris, 1838), p. 360.

3. J.H. Clapham, *The Economic Development of France and Germany, 1815-1914*, 4th edn, (London, 1961), p. 11.

4. M. Agulhon, *La République au Village* (Paris, 1970), p. 87.

5. Anon., 'Département du Drôme', *AS*, vol. 2 (1802), p. 390; Anon., 'De l'état de l'agriculture dans le département de l'Aveyron', *AAF*, vol. 2 (1822), p. 167.

6. M. Allaud, *Mémoire sur le reboisement et la conservation des bois et forêts de la France* (Limoges, 1845), p. 9.

7. J.P. Quénot, *Statistique du département de la Charente* (Paris, 1818), p. 423.

8. M. de Sainte-Colombe, 'Notice sur l'instruction publique, l'agriculture et l'industrie de l'arrondissement d'Yssingeaux', *Annales de la Société d'Agriculture, Sciences, Arts et Commerce du Puy*, vol. 3. (1828), p. 116-17.

9. Anon., 'De l'étendue, du revenu et de l'administration des forêts en France', *JAP*, vol. 2 (1838-9), p. 457.

10. B. Mugriet, 'Aperçu sur la statistique rurale du département des Landes', *AAF*, vol. 2 (1821), p. 8.

11. M. Verronnais, *Statistique historique, industrielle et commerciale du département de la Moselle* (Metz, 1844), p. 61.

12. Anon., 'De l'étendue des fôrets', p. 456; G. Huffel, *Economie forestière* (3 vols., Paris, 1904), vol. 1, p. 355.

13. Anon., 'De l'étendue des fôrets', p. 457.

14. A. de Montureux, 'Du défrichement des forêts', *JAP*, vol. 3 (1839-40), pp. 268-9.

15. M. Baudrillart, 'Mémoire sur le déboisement des montagnes', *AAF*, vol. 8. (1831), pp. 65-78.

16. H. de Villeneuve and E. Robert, 'Revue agricole de la Provence', *APAPER*, vol. 12 (1839), p. 67.

17. J.A.V. Yvart, *Excursion agronomique en Auvergne* (Paris, 1819), p. 97.

18. M. Feissat, 'Des défrichements', *APAPER*, vol. 1 (1827), p. 449.

19. M. Bosc, 'Description générale et statistique du département de l'Aude', *AAF*, vol. 6 (1819), p. 403.

20. M. Bosc, 'Projet de boisement des Basses-Alpes', *AAF*, vol. 9 (1802), p. 260.

21. O. Leclerc-Thouin, *L'Agriculture de l'ouest de la France, étudiée plus spécialement dans le département de Maine-et-Loire* (Paris, 1843), p. 397.

22. M.A. Puvis, 'Du climat et de l'agriculture du sud-est de la France', JAP, 2nd series, vol. 3 (1845-6), p. 202.

23. M. de Chambray, 'De l'agriculture et de l'industrie dans la province de Nivernais', *AAF*, vol. 13 (1834), p. 6.

24. R. Laurent, *L'Octroi de Dijon du XIXe siècle* (Paris, 1960), pp. 118-19.

25. J.A. Baudet-Lafarge, *Agriculture du département du Puy-de-Dôme* (Clermont-Ferrand, 1860), p. 256.

26. A. Clapier, 'Des bois taillis et de leur culture en Provence', *APAPER*, vol. 9 (1836), p. 410.

27. G. de Villemotte, 'Du défrichement des forêts et du boisement des terres incultes', *APAPER*, vol. 12 (1849), p. 269; Huffel, *Economie forestière*, vol. 1, p. 349.

28. de Montureux, 'Du défrichement, p. 269.

29. J.B. Martinel, 'Etat de l'agriculture dans les Ardennes', *JAP*, 2nd series, vol. 4 (1846-7), p. 11.

30. M. de Chambray, 'Culture du pin maritime en Sologne', *JAP*, 2nd series, vol. 4 (1846-7), p. 203.

31. AD Ille-et-Vilaine 37 M.

32. F. Lullin de Châteauvieux, *Voyages agronomiques en France* (Paris, 1843), p. 427.

33. R. Laurent, 'Le secteur agricole' in F. Braudel and E. Labrousse (eds.), *Histoire économique et sociale de la France* (Paris, 1976) vol. 3 (2), p. 757.

34. Anon., *Review of the Agricultural Statistics of France* (London, 1848), p. 28.

35. A. Blanqui, *Du Déboisement des montagnes* (Paris, 1846), p. 72.

36. Anon., *Review of Agricultural Statistics*, p. 28.

Fodder Resources

Grain growing was the main objective of French farming in the 1830s but for that very reason the role of livestock in the country's agricultural systems was vital since sustained crop production would have been impossible without the manure that was returned to the soil. All 51,568,800 livestock shared the function of being dung-making machines throughout their lives but they performed a variety of additional roles. Some yielded marketable products, such as milk or wool, on a regular basis but the value of others could be realised only when they had been slaughtered. In fact, many livestock changed their function through time, for example, spending their working lives as draught animals or providers of milk but ending up as sources of meat and hide. Nothing of this complexity of function was captured by the Statistique which recorded livestock numbers and provided information on sources of meat but made no mention of milk, wool or hides which formed important components of agricultural trade. Some livestock were the objects of care and attention, since the quality of their flesh and skin contributed to the price that they commanded, but most were 'the mere auxiliary of cultivation', being treated as a 'necessary evil' or 'the essential vice' without which crops could not be grown.[1] This was particularly true in the Midi, where traditional biennial rotations did not include a petite céréale for livestock feed. Throughout Provence, Languedoc and the basin of the Garonne too much land was devoted to cropping and not enough to supporting livestock and agricultural inspectors were harsh in their criticism.[2]

Fodder resources varied enormously from district to district. Clay vales, banks of watercourses and moist mountainous areas supported natural grasslands for grazing and for hay cutting, but a mere 4,198,197 ha came into this category which contributed only a modest amount of the total fodder consumed (Figure 10.1a). Livestock rearing was associated closely with arable farming through the practice of fallowing, with most of the 6,763,281 ha of fallowland being grazed during the July Monarchy (Figure 10.1b). Pâtis, communaux, landes and bruyères provided a further 10,191,076 ha of rough grazing that was subject to varying degrees of pastoral use, as well as undergoing occasional clearance for periodic cultivation. Artificial meadows covered 1,576,547 ha, being grown predominantly as part of arable rotations and providing important sources of green fodder in the Paris Basin and parts of the Midi (Figure

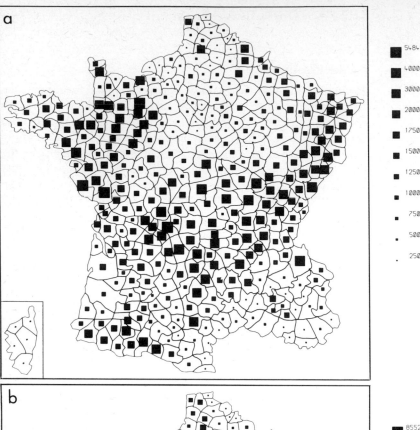

a

■	54842
■	40000
■	30000
■	20000
■	17500
■	15000
■	12500
■	10000
■	7500
·	5000
·	2500
	5

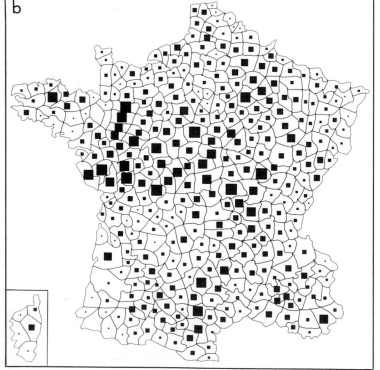

b

■	85522
■	50000
■	45000
■	40000
■	35000
■	30000
■	25000
■	20000
■	15000
·	10000
·	5000
	0

Figure 10.1: Surfaces
under (a) Natural
Grassland and (b)
Fallows (ha)

10.2). Mature grain, unripe stalks and even husks of some arable crops were fed to livestock and hence allowance should be made for these when determining how much land provided animal feed. Taking the proportion of each arable crop devoted to animal feed in 1835 as a guideline, roughly 3,000,000 ha of ploughland appear to have been used to produce livestock feed, but the list did not end there since animals were also fed on straw, potatoes, chestnuts, acorns and many types of vegetable waste.[3] The Statistique affords a very partial view of such resources; for example, a footnote stated that 607,158 qx of straw were consumed in Paris each year but no indication was given for any other part of France.[4]

Considerable attention was paid in the agricultural enquiry to natural grasslands and to the hay that they produced but much less information was given on fallows, artificial meadows or rough grazing. None the less, these fodder resources were vitally important. The compiler of the *Annales Agricoles du Département de l'Aisne* went so far as to ignore natural grasslands in his discussion of animal husbandry but concentrated on three other types of pastoral resource.[5] First, in some parts of Nord livestock were fattened intensively on root crops and were kept indoors for most of the year. Their manure was collected carefully and was applied to the soil as needed. Around Lille livestock husbandry had become a real industry. Other parts of Nord contained enclosed *pâturages* that were ploughed, seeded and used exclusively for grazing, hence they were quite different from the usual 'natural grassland'. Second, in Beauce, Brie and Normandy artificial meadows provided valuable fodders to supplement fallow land and important improvements in sheep raising had taken place in these pays after the 1790s. Third, came the rest of France where the meagre resources of fallows and of areas of rough grazing provided virtually the only supplies of animal feed. Although ignoring the lush grasslands of Lower Normandy and high *pelouses* in the mountains, this sketch was accurate enough. When all forms of pastoral resource are taken into account, no less than half of France was devoted to raising livestock on the eve of the railway age, but most areas yielded pathetically small amounts of fodder.

Natural grasslands covered a mere 8.29 per cent of France, although arrondissements in Lower Normandy had about one third of their area under grass and in some mountainous areas and in coastal sections of Vendée the proportion approached 20 per cent (Figure 5.1b). At the other extreme, less than 5 per cent of the Paris Basin was devoted to natural grasslands, with the proportion in Beauce falling below 2.5 per cent. Such small percentages were also typical of a vast stretch of southern France. The large arrondissements of Les Sables d'Olonne (1,070,000 qx)

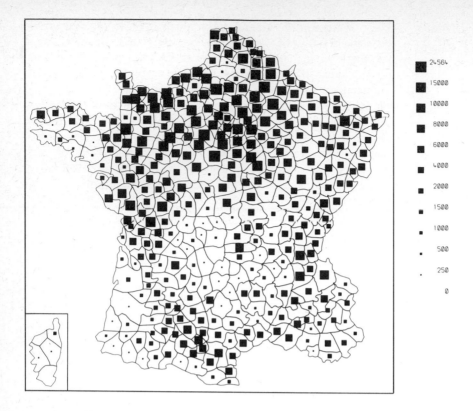

■	24564
■	15000
■	10000
■	8000
■	6000
■	4000
■	2000
■	1500
▪	1000
▪	500
.	250
	0

and Fontenay (1,010,000 qx) in Vendée each generated about one per cent of the national hay yield, being followed closely by Tulle (1,020,000 qx) in Corrèze, Saint-Lô (1,010,000 qx) and Valognes (980,000 qx) in Manche, and Aurillac (960,000 qx) in Cantal. By contrast, many predominantly arable arrondissements in Provence, Languedoc, Champagne and the Ile-de-France produced less than 100,000 qx apiece from their rare waterside meadows.

Variations in environmental conditions and management techniques led to substantial differences in the amounts of hay produced from each ha of meadow each year (Figure 10.3a). Many natural meadows in Charente and doubtless elsewhere might be described as 'stagnant marsh' or 'covered with savage waters'.[6] Efficient drainage was required if their yields were to be raised. By contrast, meadow irrigation had been practised for centuries in the Midi and was emulated with success in Haute-Vienne and parts of the Pyrenees.[7] The lowest hay yields were recorded in Tarn-et-Garonne (10.00 qx/ha) and around Brignoles (10.92) in Var, while the highest values occurred in the contiguous arrondissements of Aix (75.87) and Marseilles

(80.04). This contrast was striking in the extreme, with irrigation and a second growth of grass producing the high yields in Bouches-du-Rhône (70 to 80), Vaucluse (45 to 65) and the arrondissement of Tarbes (49.99) which were double the national average (25.06), but the amount of meadowland in these southern districts was slight. Progressive farming areas in Beauce, Upper Normandy and Flanders generated yields in excess of 40 qx/ha, as did various types of coastal environment in Charente-Inférieure, Finistère and Manche. The grassy mountains of Cantal yielded just over 30 qx/ha each year and a similar level of productivity was found in Calvados, Armor, the environs of Paris and parts of Brie. In spite of their significance in the land-use pattern the natural grasslands of Perche, Nivernais, Franche-Comté and the greater part of the Pyrenees yielded well under 20 qx/ha.

In many arrondissements all hay was consumed locally and over many remaining parts of the country the difference between production and consumption was slight, with most shortages being made good by supplies from neighbouring arrondissements. For example, Lyons was able to draw on the fodder resources of the pays of Bresse and Dauphiné nearby. Eighteen arrondissements recorded hay surpluses in excess of 100,000 qx apiece. Seine consumed 415,000 qx more than it produced each year and, although no surpluses were declared in Seine-et-Oise or Eure-et-Loir, it would be possible to satisfy Parisian requirements from the combined surpluses of Seine-et-Marne, Oise, Aube and Yonne which exceeded 550,000 qx, although periods of very high or very low water in the rivers of the eastern Paris Basin caused interruptions in supply. Arrondissements in Lorraine could more than compensate for shortages in Alsace (eg. Strasbourg −137,000 qx, Saverne −95,000 qx), while surpluses from Burgundy could meet the requirements of Franche-Comté (eg. Gray −116,000 qx). Surpluses from the well-watered valleys of Tarbes (+223,391 qx) and Bagnères (+176,597 qx) more than satisfied the needs of the Pays de l'Adour, where the arrondissements of Bayonne and Oléron consumed 98,000 qx more than they produced. With the exception of movements to Paris, Lower Alsace, Franche-Comté and the Pays de l'Adour, large quantities of hay did not appear to enter into domestic trade over the greater part of France in the 1830s.

Hay prices were surprisingly uniform over large parts of the country, with the usual cost being 4 to 5 F/quintal (Figure 10.3b). In five areas, where production balanced consumption or there were surpluses, prices fell to 3 F, descending even further around Saverne (2.80) and Brest (2.00). The mountainous grasslands of Cantal and Lozère formed the first of the five areas,

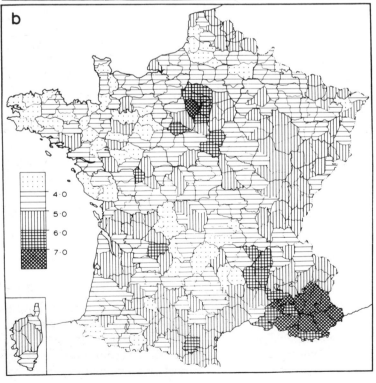

Figure 10.3: Hay
(a) Annual Yield
(quintaux/ha) (b) Price
(F/quintal)

a

20
25
30
35
40

b

4·0
5·0
6·0
7·0

174

with other low-cost areas comprising Haute-Vienne, many parts of Brittany, the Armorican marchlands and Somme. Prices rose to 6 F/quintal in the capital and its environs, which had to be supplied from pays to the immediate east. Indeed, it was claimed that Parisian hay demands inflated prices in Brie and other parts of Seine-et-Marne and thereby prevented stock rearing from developing as fully as it might have done but it was in Provence that the highest prices were paid, especially around Sisteron (9.65 F), Forcalquier (9.85) and Aix (10.00).[8] Surprisingly, each of these arrondissements generated surpluses and displayed some of the highest yields in France but presumably prices were high in order to cover irrigation costs. Once again, Provence was the most expensive area in France, with a secondary pole of high prices around Paris.

Price (F/quintal) multiplied by production (qx/ha) produces a measure of the financial yield of natural grassland. Nationwide this amounted to 110.20 F/ha but there were very substantial deviations from the mean. Values were highest for the irrigated meadows of the Midi where hay supplies were short, with sums of roughly four to six times the national average characterising the arrondissements of Arles (423), Avignon (456), Marseilles (600) and Aix (758). Nord, Upper Normandy and the environs of Paris registered values that were double or just under double the national average. By contrast, very low values were recorded over much of Maine, Perche, Franche-Comté, the Causses and the Pays de l'Adour.

Such detail is not given for other sources of fodder. Pâtis, communaux, landes and bruyères represented one-fifth of the land surface and were extensive in the Massif Central, Alps, Landes and Armorica even though much rough grazing had been cleared for arable use after the Revolution. For example, the shortage of fodder as a result of défrichement was declared to be 'a real calamity' in Moselle which meant that all livestock were 'excessively thin'.[9] Artificial meadows needed to be grown but for this solution to be effective property boundaries would have to be respected and the 'disastrous custom' of parcours be abolished. Identical problems were experienced in many other parts of the country. Pâtis were not recorded by arrondissement but their distribution must have been virtually identical to that for 'other' land uses shown on Figure 5.2b.

Fallows covered 6,763,281 ha and produced fodder that was valued at 92,285,902 F or 13.65 F/ha, which was roughly half as much again as the average ha of rough grazing. High values exceeding 20 F/ha and even 30 F/ha were associated with heavily-stocked départements containing large cities, for example, Nord, Pas-de-Calais, Seine-Inférieure and the environs of Lyons, and also a wide stretch of south-western France over which biennial rotations operated and fallow made a truly crucial contribution to livestock feed.

The range of fodders recognised as artificial meadows occupied 1,576,547 ha and produced animal feed valued at 203,765,169 F. This averaged 129.25 F/ha, being slightly superior to that for natural grassland and in a completely different league from the miserably low productivity of rough grazing and fallow. Output was valued most highly (over 160 F/ha) in the Ile-de-France, Nord, Pas-de-Calais, Upper Normandy, north-eastern France and Provence. Values of less than 80 F/ha were typical of western France where artificial meadows were not widespread. Variations in the price of fodders derived from artificial meadows of course displayed a very similar pattern, rising to more than 5.5 F/qx in parts of the Ile-de-France, Provence and the north but falling below 3.5 F/ha in sections of the north-west.

Livestock

France supported no fewer than 32,151,400 sheep in the mid-1830s (Table 10.1) with a roughly crescent-shaped stretch of high densities running from the Ardennes, through the inner Paris Basin, Berry, Limousin and Poitou to the Causses, Languedoc and the Lower Rhône. Adopting all forms of land except woodland as the basis for calculation, densities exceeded 200 head/100 ha in parts of Dordogne (Nontron 209), the northern Paris Basin (Soissons 212), the Camargue (Arles 228) and Berry (Issoudun 248). Artificial meadows provided valuable fodder in parts of the Paris Basin but fallows and rough grazing were the major sources in most sheeplands. Densities fell away sharply to both east and west of the crescent with less than 25 sheep/100 ha being supported in eastern arrondissements from Sarrebourg to Trevoux, in Maine and much of Brittany.

In 1787 merinos had been introduced on the Rambouillet experimental farm founded two years earlier by Louis XVI and in

Table 10.1: Livestock

Cattle	9,936,500
Sheep	32,151,400
Pigs	4,910,700
Goats	964,300
Horses	2,818,500
Mules	373,800
Asses	413,500
	51,568,700

the following half century important improvements were made to sheep bred in the Paris Basin which were nourished increasingly on a diet of fodders from artificial meadows.[10] It was claimed that Beauce had more merinos than any other part of France, while in Champagne the number of sheep doubled between 1773 and 1840 largely because of the successful raising of Spanish merinos which had been popularised by Lafayette among others. New forms of stock-keeping evolved in the Paris Basin in the early nineteenth century but the process of défrichement withdrew traditional sources of sheep fodder throughout the country. In pays such as Brie, artificial meadows formed an excellent replacement but in Puy-de-Dôme more intensive cultivation in the Limagnes had reduced the number of sheep that could be supported in the lowland and in Aude it was claimed that sheep numbers had fallen from 1,000,000 at the Revolution to 595,700 in 1812 following extensive défrichement in the garrigues.[11] Nevertheless the quality of Corbières sheep was reported to be good and the same was true for Larzac and Saint-Affrique sheep further to the north, where local breeds had been crossed with merinos. The resulting strains produced not only fine fleeces that were used in the textile workshops of Carcassonne but also meat of a high quality and milk that was used to produce Roquefort cheese for distribution to Bordeaux, towns in the Midi and even Paris.

Such praise was rare indeed since most contemporaries condemned the way that sheep husbandry was neglected. The practice of transhumance that was widespread in the south was criticised as a 'vicious practice' which represented 'the remains of nomadic life'.[12] Roughly 1,000,000 sheep wintered on and around the Rhône delta each year with a large proportion moving to upland pastures for four or five months each summer. Three-quarters of the sheep in Bouches-du-Rhône were transhumants, with only 150,000 spending the summer in the département, especially in communes to the north and east of Aix.[13] The remaining 450,000 moved into the Alps from early June to late September, thereby depriving the lowlands of large quantities of manure that were needed for olive trees and other forms of intensive cultivation. Identical complaints surrounded the movement of sheep from the Crau, the Camargue and the lowlands of Languedoc and Provence into the southern Alps and high pastures in the southern Massif Central. Short-distance transhumance also took place, as in the mountains of Var where sheep moved downslope in December until late April and then returned to higher grazing grounds during the summer months. Hillsides were shorn clean of vegetation as transhumant flocks devoured whatever might serve as fodder and woodcutters removed the little timber that remained. Similar movements produced identical responses in the Aubrac, where 30,000 to 40,000 sheep from Aveyron, Lozère, Cantal, Gard and Hérault spent the summer, and in Vivarais which was invaded by 100,000 sheep each June.[14]

The 9,936,500 cattle recorded in the Statistique were distributed in a very different way to the nation's sheep, with large numbers being found in a fragmented ring around the Paris Basin, extending from Vendée through Armorica, Normandy, Flanders, Alsace, the Vosges, Franche-Comté, and the Bresse to Dauphiné and Forez (Figure 10.4a). Densities exceeded 50 cattle/100 ha in fourteen arrondissements, with peak values in the arrondissements of Paris (108), Weissembourg (71), Lille (70) and Saint-Dié (69) and more than 30 head/100 ha being widespread throughout the ring. With the exception of the environs of Paris, and Mantes and the Pays de Bray such densities are not encountered in the Paris Basin, where 10 to 20 cattle/100 ha was much more typical. Mediterranean France was virtually devoid of cattle, with less than 2 head/100 ha being recorded in fifteen arrondissements. Between 20 and 30 cattle/100 ha was typical of the pays charentais and almost all of the Massif Central. Only in areas of mountain grassland, such as Cantal, did the figures reach 30 to 40/100 ha. Similar densities were recorded in seven arrondissements in the middle Garonne and the Pyrenean foothills.

Very different sets of pastoral resources were involved in raising sheep and cattle, for whereas the distribution of sheep reflected fallow land and to a lesser extent artificial meadows, cattle raising was concentrated in the permanent grasslands of Armorica, Normandy, Flanders, the north-east and parts of the Massif Central. Hay cost less than 5 F/quintal and often less than 4 F over the greater part of Armorica and local supplies could satisfy demands in almost all western arrondissements. Between 5 and 6 F/quintal was the usual price of hay throughout Flanders but demands were satisfied by local supplies. Similar prices were encountered in the north-east as a whole, regional supplies were sufficient to meet needs but local movements of hay were necessary between areas of surplus and shortage. Cattle were rare in Provence where hay prices were high (over 7 F/quintal) and, with the exception of the département of Seine, cattle densities were low in the Ile-de-France where hay prices were over 6 F/quintal.

Natural meadows formed the major fodder resource for cattle but the better areas of rough grazing also played their part, especially in moist mountainous areas. Cattle were also involved in transhumance with, for example, 15,000 to 20,000 cattle from Aveyron, Lozère and Cantal spending part of the summer in the mountains of Aubrac.[15] Beet and other fodder roots were also

Figure 10.4: (a) Density
of Cattle/100 ha
(b) Ratio of Sheep to
Each Head of Cattle

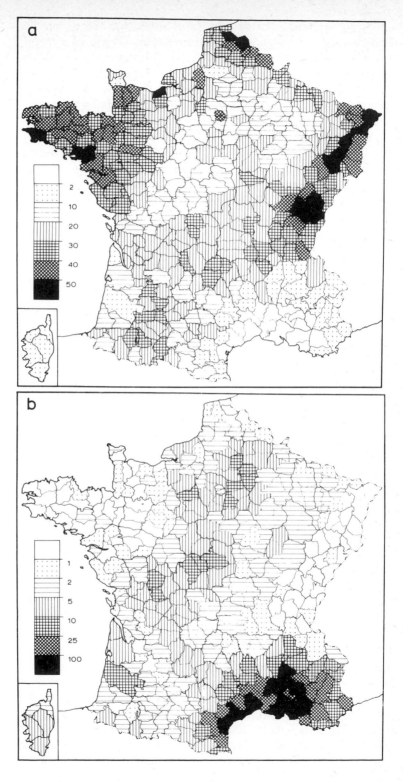

179

highly important as cattle feed but it seems that sheep were the main beneficiaries from artificial meadows. The grassy mountains of the Massif Central formed *pays naisseurs* par excellence and de Parieu described cattle as the 'true inhabitants' of the highlands of the Auvergne. [16] Stock born in Cantal and Limousin were despatched to Normandy, Brittany, the Marche, Nivernais and Charolais as work animals and also for fattening prior to being butchered for the Paris market and, in the case of the last two pays, for Lyons. Cattle born in the mountains of Mézenc were despatched as work animals to the surrounding départements of Haute-Loire, Lozère, Drôme, Isère and Ardèche. The latter département, in turn, sent work animals to the Rhône valley and to Forez, as well as fattening bullocks and calves for the butchers of Lyons. Movements such as these proved interesting examples of a response to market demands in regions that in many other respects remained largely inward-looking during the July Monarchy. Cattle moved on their own feet from the uplands of France but few commodities were carried back. Stock raised in the mountains of Franche-Comté and Switzerland were fattened in Haute-Saône prior to being sent to Nord, Lorraine and Alsace, although the latter region was also supplied with fatstock from across the Rhine. [17] Pulp from sugar beet provided a useful source of feed in Aisne and Nord for livestock purchased in Nivernais where Durhams had been introduced successfully in the mid-1820s. Thin cattle were sent from Lower Maine and Poitou for fattening in Perche, Cotentin, Bessin and the Pays d'Auge prior to being sent to Beauce, Brie and other parts of the Ile-de-France as the final stage before the market and abattoirs of Sceaux and Poissy.

Information in the Statistique on *bovins* was disaggregated into *taureaux* (399,026), *boeufs* (1,968,838), *vaches* (5,501,825) and *veaux* (2,066,849). Boeufs performed the dual role of fatstock and draught animals and their distribution bore some similarities to that for all forms of cattle with concentrations in the north-west and north-east but very low densities in the Midi méditerranéen. No distinction was drawn between dairy cows and cows being fattened for butchery but the 5,501,825 vaches recorded in the Statistique made up just over one half of all cattle. The ring of cattlelands emerges clearly, with virtually every arrondissement in Brittany, Upper Normandy, Nord, Pas-de-Calais, the north-east, Dauphiné and the heart of the Massif Central supporting more than 20 cows/100 ha. Densities were slightly less in Lower Normandy and Lorraine but exceeded 35/100 ha around Montfort (Ille-et-Vilaine), in the Vosges, interior Flanders (eg. 58/100 ha in Lille arrondissement) and the environs of Paris (eg. 105/100 ha in Paris arrondissement). Fodder resources as diverse as the mountain grasslands of the

Vosges, the enclosed pâturages of Flanders and the suburban cowsheds of Lille and the capital contributed to these very intense concentrations. [18] The quality and breeds of the cows that were kept varied in response to available fodder resources. Thus small Morvan cows, which had low milk-yields but required little fodder, were kept in areas of mediocre farming around Fontainebleau and Melun in Seine-et-Marne. [19] By contrast, much larger Normandy cows which gave good milk yields were kept in the richer pays of Brie and Multien, while in the environs of Meaux cows from Flanders and Picardy were preferred which, it was claimed, gave yields that were far superior to those from the best cows from Cotentin. Only when cows had fulfilled their role as producers of calves and milk were they sold off as meat for the Paris market.

The distribution of calves bore a close similarity to that for cows, emphasising Armorica, Flanders and the north-east, with the highest densities around Quimper (18/100 ha), Hazebrouck (19), Weissembourg (20) and Strasbourg (15). Calves were fattened for the Paris market in Vexin, Brie and Beauce, being sold either direct to the capital or via markets such as Pontoise, Coulommiers and Sens. They were fed on milk and barley flour and made few direct demands on fodder resources. Calf-fattening was one of the most productive branches of agriculture in Brie in 1835, having acquired greater importance since 1830. [20]

The contrast between sheeplands and cattlelands is summarised in Figure 10.4b which indicates the number of sheep for each head of cattle. Arrondissements in Armorica, Flanders and the north-east contained more cattle than sheep but in all other parts of France the reverse was true, with sheep outnumbering cattle by more than one hundred to one in ten arrondissements in the Narbonne-Montélimar-Aix triangle and ratios of more than ten to one characterising the whole of the Midi méditerranéen. Ratios declined gradually northwards into the Massif Central but a much sharper break was in evidence between the sheeplands of Drôme and the cattle country of Isère. Only in Berry, Upper Poitou, Soissonnais and parts of the Landes were there more than ten sheep for every head of cattle.

The number of pigs raised in France had increased substantially since the beginning of the nineteenth century to reach 4,910,700 in the mid-1830s. Pigs were fed on a wide range of substances including vegetable peelings, chestnuts, acorns, flour, cabbage, whey and even the litter from silkworms. [21] Pigmeat had a high calorie-to-weight ratio and made a valuable contribution to otherwise largely vegetarian diets in many parts of the French countryside. The pig came to the rescue in many areas with high population densities surviving on a limited range of agricultural resources, but living standards improved as the century progressed and the number of pigs declined as greater quantities of beef were eaten. [22] During the July Monarchy that downward trend had not yet begun and the highest densities of pigs (over 25/100 ha) were in regions where the potato had been accepted early, for example in the north-east and Périgord, rising to 46/100 ha in the environs of Nontron (Dordogne), and being very high around Strasbourg (37) and Weissembourg (38) in Bas-Rhin and in Louhans arrondissement (36) in Saône-et-Loire. The central Pyrenees and Anjou, together with Pas-de-Calais, formed other foci of potato growing and pig raising. Few pigs were raised in relatively empty areas, such as the Alps, Landes and sections of the Paris Basin, or along the Mediterranean littoral. As well as providing a vital supplement to the local diet, pig rearing formed an element of agricultural trade in a crescent of areas on the southern margin of the Massif Central.

Not far short of 1,000,000 goats eked a living from France's rough grazing land in the 1830s. Not surprisingly they were virtually absent from predominantly arable areas such as the Paris Basin and the basin of the Garonne but they were numerous in regions with dissected terrain and rough vegetation such as the Massif Central, Provence and Corsica. They were unquestionably a mixed blessing serving as 'the poor man's cow' to provide milk and meat and enhance otherwise meagre diets but also causing serious damage in fragile environments. [23] In the Auvergne every food shortage encouraged the poor to keep more goats and the neighbouring département of Corrèze was described as 'covered with goats' which caused serious damage to young trees and hedges. [24]

The remaining livestock recorded in the Statistique were draught animals and beasts of burden which, one suspects, were not entirely agricultural. There were 2,818,500 horses mainly distributed in an arc of horse-ploughing territory, with more than 10 horses/100 ha, which extended from Finistère through the northern Paris Basin to Nord, where all ploughing was done by horses, and on to Haut-Rhin. Densities exceeded 20/100 ha in seventeen arrondissements that contained large cities (eg. Seine 98/100 ha) or garrison towns in which cavalry was quartered. No less than half of the national cavalry was stationed in north-eastern France where very large quantities of hay and straw were consumed. The urbanised arrondissement of Marseilles (39/100 ha) was the only southern area to support more than 20 horses/100 ha and, with very few exceptions such as the ploughlands of Dauphiné, all arrondissements south of the Loire contained fewer than 5 horses/100 ha, with densities falling below 2.5/100 ha throughout the Alps and the Massif Central. Horses were generally larger and heavier in the northern third of the country than the stunted beasts that could be observed in the Garonne valley and elsewhere in the south where oxen normally pulled ploughs.

Small numbers of mules (373,800) and asses (413,500) served as beasts of burden and draught animals in southern and western France. Only in Provence, Languedoc, Périgord and Poitou were there more than 2 mules/100 ha, with virtually none in Brittany and the north-east. They were used for ploughing in the Vendée and parts of Aquitaine and Languedoc, being brought into the latter region from Poitou, Savoy and Upper Dauphiné where they were particularly esteemed as beasts of burden because of their surefootedness. Asses were found in roughly similar areas, with the addition of the Ile-de-France and Champagne where they were particularly useful on small holdings.[25]

In 1814 Morris Birkbeck expressed his disgust at the primitive state of livestock rearing in many regions of France.

> A very few half-starved sheep are kept to pick over the constantly recurring barren fallows, often accompanied by three or four long-legged hogs. On the borders and out of the way corners you may see a cow or two with an attendant. But there appears so little for any of the animals to eat that you wonder how they are supported.[26]

His words would seem to be almost equally apposite to describe conditions during the July Monarchy since cultivation of artificial meadows and intensive raising of stock, as in Normandy and Flanders, remained very much exceptions to the rule. In Ysabeau's opinion France was capable of producing four or five times more animal feed that she generated in the 1830s if fallows and rough grazing were to be replaced by fodder crops and permanent grass.[27] Dezeimeris predicted that cereal yields would decline, unless more livestock were kept and more manure were applied to the soil, and that France's rapidly growing population would be faced with ruin.[28] Already many of Louis-Philippe's subjects were eating scarcely enough to keep body and soul together.

Meat Prices

In response to the national dearth of fodder meat was a rare commodity which appeared infrequently on the tables of the poor and was consumed in very small quantities across many sections of the countryside, although it appeared more prominently in the diets of townsfolk, especially those in the more privileged strata of society. Underlying these general statements were important regional differences in the availability of various types of meat because of the basic facts of animal husbandry.

For example, cattle were rare in the Midi whereas sheep were numerous and this fact conditioned the geography of meat consumption in the south. The ratio between sheep and cattle was profoundly different north of the Loire and meat supplies varied accordingly. Pig raising was widespread in the countryside and especially in the north-east where pigmeat consumption was particularly pronounced. Information on butchery and meat prices reflect the geography of urban consumption rather than of animal husbandry since meat was transported on the hoof from pays that were relatively well endowed with fodder resources to towns and cities whose inhabitants could afford to purchase it.

Some 13,518,415 head of livestock were reported as being slaughtered each year, with sheep (5,704,381), pigs (3,957,387), cattle (3,699,229) and goats (157,418) in descending order of importance. Corsica accounted for 22,290 of the goats slaughtered and a further 16,540 were butchered in Var but numbers were very small in every other part of France. As the largest and most affluent demand centre Paris headed the list of livestock consumption, with no fewer than 236,260 cattle, 504,610 sheep and 117,580 pigs being butchered in the capital's abattoirs. The neighbouring départements of Seine-et-Oise and Seine-et-Marne were the main suppliers of calves, sheep and barren dairy cows, with important numbers of sheep also coming from more distant pays including Bourbonnais, Berry, the Ardennes, the Sologne, the Langres plateau, Normandy and Flanders. Fat cattle were despatched in great quantities from Lower Normandy, Maine and Nivernais, as the figures on livestock sold in 1843 at the Parisian markets of Sceaux, Poissy, La Chapelle, Bernadins and Halle-aux-Vins demonstrate most clearly. [29]

The dominance of Paris was absolute in terms of meat consumption but beyond that fact there were important spatial variations for the three main types of livestock. Large numbers of cattle were slaughtered in Lyons and Bordeaux, which were respectively the second and fourth most populous cities, but the pattern of beef butchery emphasised the northern half of France and especially the north-west, with the arrondissements of Rennes, Brest, Rouen and Nantes among the leading eight butchery centres. Mutton was the main form of meat eaten in the cities of the Mediterranean south which accounted for five of the eight leading centres of sheep slaughter, with Marseilles, France's third largest city, coming in fourth place. By contrast, the cities of north-eastern France formed the main focus of pig butchery, with Strasbourg, the ninth most populous city, coming in second position. As many as 580,650 pigs were slaughtered in the six départements of Alsace and Lorraine out of a national total of 3,957,380, with roughly equal numbers being butchered in the three départements of the Ile-de-France

(201,760) and in Nord and Pas-de-Calais (191,620) which contained similar numbers of consumers. Animal slaughter was of limited importance away from the main cities since livestock rearers in pastoral pays produced animals for sale and fodder resources in many other areas were too sparse for large numbers of livestock to be supported. Hence meat rarely entered the diet of many countryfolk. Beef, or 'butcher's meat' as it was called, was virtually unknown outside towns where it was 'the food of the rich', however, the flesh of home-raised pigs provided a small but valuable source of animal protein in predominantly vegetarian regimes in the countryside and for poorer sections of urban society.

The inhabitants of villages and small towns relied on local supplies for the small quantities of meat that they consumed each year. They were essentially beyond the commercial circuits of production and hence the prices they paid were relatively low. Much higher prices had to be paid for meat in the major cities that were supplied from pastoral areas near and far and where meat consumers had to bear additional costs associated with the transport, marketing and butchering of animals.[30] For these reasons there was an underlying similarity in the broad spatial arrangements of prices for each of the three main types of meat.

Beef prices ranged from 5.5 F/kg around Marvejols (Lozère) and Lectoure (Gers), where very little butcher's meat was eaten, to exactly double that amount in Beauvaisis and Eure where wheat prices were also high (Figure 10.5a). In Paris and in the towns of the Ile-de-France beef cost 9.5 F/kg and prices in excess of 9.0 F/kg characterised Strasbourg, the remainder of Gironde and the few remaining sections of the northern Paris Basin. Price levels declined rapidly beyond the Ile-de-France, Upper Normandy, Picardy and Flanders with a remarkably sharp gradient between adjacent arrondissements in Eure (11.0) and Orne (7.0). The pays of the Rhône and Saône, where beef cost 8.0 F/kg, formed the only major zone in which prices exceeded 6 to 7.0 F/kg which was normal over much of central and south-western France where few cattle were slaughtered and very little beef was eaten. Mutton prices were greatest around Arles (13.0), with the capital occupying second position (11.5) and a kg of mutton costing 10.0 F around Bordeaux and Toulouse, throughout lower Provence and in Eure, Boulonnais and the Flemish coastal plain (Figure 10.5b). Prices exceeded 8.5 F/kg in much of the Garonne valley, Languedoc, Upper Normandy, the Ile-de-France, Flanders and Artois which formed the foci of mutton consumption. Prices were substantially lower elsewhere falling to 6.5 F/kg or even less in much of the Massif Central and Armorica where little mutton was eaten and only local supplies

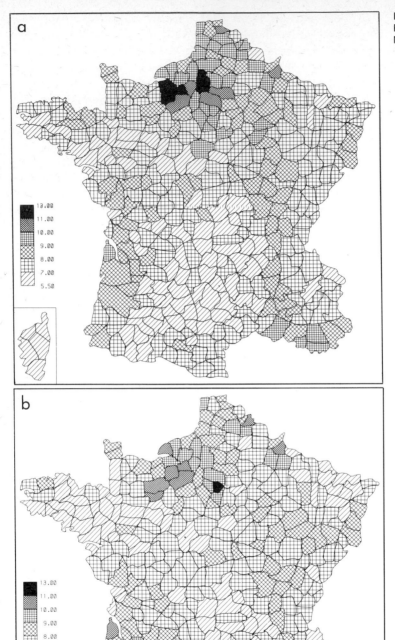

Figure 10.5: Average Price (a) of Beef (b) of Mutton (F/kg)

Legend (a):
13.00
11.00
10.00
9.00
8.00
7.00
5.50

Legend (b):
13.00
11.00
10.00
9.00
8.00
7.00
4.50

were involved. Pigmeat prices also conformed to the same general pattern with a peak price of 12.0 F/kg around Louviers in Eure, where all forms of meat and cereals were expensive; around Valenciennes in the densely populated, industrialising interior of Flanders, which was a powerful demand centre for all types of food; and around Arles where mutton (the major local source of meat) was also expensive. Pigmeat was moderately priced in Alsace and Lorraine (8 to 9.0 F), which were well provided with local supplies, and prices were substantially lower across widely dispersed arrondissements in western and central France where pigs were raised to provide some animal protein.

Frequent meat-eating was a privilege that only the rich could afford during the July Monarchy and the record of livestock slaughter and meat prices has served to identify the towns, cities and more progressive countrysides which contained sizeable proportions of relatively affluent households as well as supporting their share of urban and rural poor. It has also delineated the 'other' France where a harsher existence and a frugal dietary regime was the lot of the greater part of the population.

Notes

1. P. Bozon, *La Vie rurale en Vivarais* (Clermont-Ferrand, 1961), p. 96; A. Puvis, 'Du climat et de l'agriculture du sud-est de la France', *JAP*, 2nd series, vol. 3 (1845-6), p. 255; J.E. Dezeimeris, 'Plan d'améliorations agricoles pour le département de la Dordogne', *Annales Agricoles et Littéraires de la Dordogne*, vol. 1 (1840), p. 285.

2. Inspecteurs de l'Agriculture, *Agriculture française: Tarn* (Paris, 1845), p. 375; Inspecteurs de l'Agriculture, *Agriculture française: Haute-Garonne* (Paris, 1843), p. 104.

3. Ministère des Travaux Publics, de l'Agriculture et du Commerce, *Archives Statistiques* (Paris, 1837).

4. Ministère de l'Agriculture, *Statistique de la France: Agriculture* (4 vols., Paris, 1840, 1842), vol. 2, 1840, p. 183.

5. Anon., 'Réponse aux questions ministérielles sur la situation de l'agriculture dans le département de l'Aisne', *Annales Agricoles du Département de l'Aisne*, vol. 7 (1835), p. 51.

6. J.P. Quénot, *Statistique du département de la Charente* (Paris, 1818), p. 406.

7. M. Bugeaud de la Piconerie, 'Irrigations dans le département de la Haute-Vienne', *AAF*, vol. 5 (1821), pp. 93-112; P. Boulinière, *Itinéraire déscriptif et pittoresque des Hautes-Pyrénées françaises* (3 vols., Paris, 1825), vol. 3, p. 282.

8. AD Seine-et-Marne M 9279. Etat des récoltes et prix des fourrages. Document from prefect dated 28 October 1841.

9. M. Bottin, Etat de l'agriculture dans le canton de Mars la Tour, *Mémoires de la Société Libre d'Agriculture pour le département de la Moselle*, vol. 1 (1803), p. 33.

10. Mme Romieu, *Des Paysans et de l'agriculture en France au XIXe siècle* (Paris, 1865), p. 255.

11. AD Seine-et-Marne M 7323. 'Situation et progrès de l'agriculture, 1835; F. Jusserand, 'Statistique agricole de la commune de Vensat, mémoire pour servir à une description de l'agriculture de la Limagne', *Bulletin Agricole du Puy-de-*

Dôme, vol. 1 (1841-2), p. 75; M. Bosc, 'Description générale et statistique du département de l'Aude', *AAF*, vol. 6 (1819), p. 405.

12. H. de Villeneuve and E. Robert, 'Revue agricole de la Provence', *APAPER*, vol. 12 (1839), p. 64.

13. M. de Belleval, 'Réflexions sur la transhumance des troupeaux en Provence', *APAPER*, vol. 20 (1847), p. 54.

14. A. Meynier, *A travers le Massif Central: Ségala, Levezou, Châtaigneraie* (Aurillac, 1931), p. 117; Bozon, *Vivarais*, pp. 100-1.

15. G. de Buzareinges, 'Mémoire sur les montagnes d'Aubrac', *AAF*, vol. 12 (1833), p. 34.

16. F.E. de Parieu, *Essai sur la statistique agricole du département du Cantal*, 2nd edn. (Paris, 1864), p. 43.

17. J.A. Marc, 'Aperçu des ressources agricoles du département de la Haute-Saône', *Mémoires de la Sociéte d'Agriculture, Sciences, Commerce et Arts du Département de la Haute-Saône*, vol. 3 (1812), p. 236; M. Wohlfart, 'Agriculture de l'Alsace', *JAP*, vol. 5 (1841-2), p. 251.

18. J. Cordier, *Mémoire sur l'agriculture de la Flandre française et sur l'économie rurale* (Paris, 1823), p. 418; C. de Gourcy, *Notes sur l'agriculture des départements du Nord et du Pas-de-Calais* (Paris, 1847), p. 11.

19. A. Rayer, *Etude sur l'économie rurale du département de Seine-et-Marne* (Paris, 1895), p. 13.

20. AD Seine-et-Marne. M. 7323. Rapport spécial sur la situation de l'agriculture dans l'arrondissement de Coulommiers en 1835.

21. Bozon, *Vivarais*, pp. 110-11; R. Béteille, *Les Aveyronnais* (Poitiers, 1974), p. 30.

22. M. Block, *Statistique de la France* (2 vols., Paris, 1876), vol. 2, p. 96.

23. M. Lange, 'Rapport sur l'ouvrage de M. Yvart ayant pour titre Excursion agronomique en Auvergne', *Mémoires de la Société d'Agriculture et du Commerce de Caen*, vol. 3 (1830), p. 75.

24. A. Firmigier, 'Essai de statistique du département de la Corrèze', *AS*, vol. 4 (1802), p. 178.

25. M. Block, *Statistique de la France*, vol. 2, p. 84.

26. M. Birkbeck, *Notes on a Journey through France* (London, 1814), p. 110.

27. A. Ysabeau, 'De la production et du commerce des fourrages en France', *JAP*, vol. 5 (1841-2), p. 120.

28. J.E. Dezeimeris, 'Des moyens d'améliorer l'agriculture en France', *JAP*, vol. 5 (1841-2), p. 576.

29. A. Husson, *Les Consommations de Paris* (Paris, 1856); J. Vidalenc, 'L'approvisionnement de Paris en viande sous l'ancien régime', *RHES*, vol. 30 (1952), pp. 116-32; J. Vidalenc, *Le Peuple des campagnes: la société française de 1815 à 1848* (Paris, 1970), pp. 375-8.

30. R. Laurent, *L'Octroi de Dijon au XIXe siècle* (Paris, 1960), p. 100.

11

Sources and Methods

The mosaic of agricultural production during the early years of the July Monarchy was mirrored by an equally complex pattern of food consumption as 33,500,000 people survived on the cereals, 'special crops' and livestock products that France generated. In a context of virtual self-sufficiency in basic foodstuffs but limited opportunities to transport commodities over long distances, spatial variations in the quantity and quality of diet provided an eloquent commentary on the diversity of agricultural systems, on their efficiency as a means of food production, and on the relative well-being of the consuming population. For most Frenchmen eating consisted of 'a lifetime of consuming bread . . . still more bread and gruels'.[1] In many parts of the countryside meat was a rarity that was eaten perhaps only once or twice a week and then it was most likely to be some cured or salted form of pigmeat. 'Butcher's meat' was 'still considered to be an object of luxury and was not regarded as indispensable to life or even to health'.[2] In many villages it was eaten only on the local patron saint's day, when the harvest was safely home, or if an unexpected shortage of winter fodder required cattle to be slaughtered and eaten or salted for later use. Fish, poultry and game were consumed in greatly varying quantities depending on the local environment and contributed some animal protein to what was a largely meatless diet for the majority of Louis-Philippe's subjects.

By comparison with earlier centuries, wheat figured more prominently in the human diet in the 1830s. None the less, a very wide range of sources of carbohydrate continued to be used. White bread made with pure wheaten flour was eaten by the bourgeoisie and by countryfolk in some wheat-growing areas if the grain was not despatched as a cash crop. Town dwellers and the higher orders of society elsewhere in France preferred white bread but most of the rural population were not so privileged. Contemporary observers provided vivid accounts of the varieties of black, brown and reddish bread that were made where rye, maslin, barley, buckwheat, maize or pulses formed part of the flour which, in any case, was often inadequately milled so that large quantities of bran remained.[3] Inclusion of bad grains caused further discoloration. Some grains, such as maize, buckwheat and oats, were used for making gruels and broths which incorporated skimmed milk or simply water. Buckwheat was also used for making *galettes* which could be up to a foot across and half an inch thick.[4] In some areas, including the mountains of Auvergne, rye flour was mixed with milk and cooked into a kind of omelette.[5] Potatoes provided a source of carbohydrate and appeared in soups throughout the country, whereas chestnuts formed a valuable addition to the human diet in their areas of production. Such observations were, unfortunately, expressed in non-quantifiable ways and may not be assembled to produce a coherent picture of the quantities of food being eaten throughout the country in the 1830s.

Food consumption later in the nineteenth century has been examined more rigorously using information from four types of source: decennial agricultural enquiries, statistics on food intake in cities, monographs of peasant life (prepared in the tradition of Le Play) and retrospective sections of local enquiries initiated by Lucien Febvre in the 1930s.[6] With the exception of the agricultural enquiries, each of these sources related to specific settlements or even to individual households from which much broader spatial generalisations have been hazarded. Admittedly, the erosion of regional differences in food consumption must have been a slow process, but each of these sources dates from the railway age when France was being welded gradually into a single economic unit. Historical enquiries into food consumption in earlier centuries have made ingenious use of inventories and very varied documentation to identify rather than to quantify the characteristics of local or regional diets. The net result is a great deal of knowledge about the details of dietary behaviour in some parts of France at particular periods in the past but a lack of anything other than impressionistic literary evidence for France as a whole during the July Monarchy.

As part of the great agricultural enquiry mayors were required to indicate the volume of various cereals and types of alcohol and the quantity of meat that was consumed in their communes each year. Such information was collated at the arrondissement level and related to the resident population to provide figures on average per capita consumption for each commodity. The Statistique thus offers interesting opportunities for examination and cartographic treatment, using the data either in their original form or after some kind of transformation. In fact, surprisingly little use has been made of this source in studies of food consumption. One suspects that this may be due to five types of complication.

First, many pieces of research have been unconcerned or, at best, only partially concerned with the quantities of commodities eaten. Contemporary sources often describe the range of foodstuffs consumed, sometimes even itemising the contents of meals throughout the day, but normally give no idea of quantities involved. Records of the dietary behaviour of a very limited number of households and settlements have survived for a slightly later date in the writings of Le Play but they too lack precise quantified evidence.[7] As a result the established methodology for studies of food consumption has not been a quantitative one.

Second, most of the food eaten in the 1830s was not marketed. Many rural families were self-sufficient to a greater or lesser degree and purchased only limited quantities of commodities that could not be produced on the farm.[8] The data in the Statistique, for all their apparent precision, must to a large extent have been estimates, albeit ones that were checked and modified at various levels in the administrative hierarchy. They should be regarded as suggestive rather than definitive statements.

Third, the evidence contained in the Statistique raises formidable contextual problems since it represents arithmetic means of food consumption by the 'average' inhabitant of each arrondissement, be he (or she) peasant, artisan or bourgeois, young or old. This kind of lack of information about the nature of the consumer is also encountered in many of the more literary enquiries that have been mentioned above. In each case it severely restricts the kinds of question that may be asked of the data. For example, one may not enquire too deeply whether the average diet in a particular area was 'adequate', since it is not possible to determine for whom and for the performance of which tasks it might have been more or less sufficient. It would be reasonable to assume that there were dietary differences between rich and poor and between townsmen and countryfolk in the 1830s but the Statistique does not allow such social phenomena to be investigated. However, guarded comments may be made on variations in food consumption between arrondissements that contained large cities and those that did not.

Fourth, certain elements of diet were not recorded in the Statistique. Fish, game, dairy products, fruit, edible oils and fats were the leading omissions and for that reason total diets cannot be identified. The degree of incompleteness must have varied from area to area, being perhaps least serious in northern arable regions but falling far short of reality in coastal and pastoral areas as well as in southern localities where polyculture and the basse-cour made significant contributions to the peasant diet.

Finally, the commodities were recorded in different types of unitary measure, with some expressed by volume (hl) and others by weight (kg). Many past studies have left foodstuffs in their original units of expression, regardless of differences in food value. Carbohydrates, for instance, were recorded in units of volume rather than weight even though 1.0 hl of potatoes is very different from 1.0 hl of wheat with respect to both weight and food value. In the case of carbohydrate sources two types of statistical transformation are required (from volume to weight, and from weight to calorific value) in order to allow commodities to be compared realistically. In the case of alcohol the transformation may be made directly from volume to calorific value, while meat may be converted from weight to calorific value. Such computations have not been widely attempted in previous studies but they are certainly possible with the help of accepted conversion co-efficients. As a result, the declared components of 'average' diets may be described individually and summarised in quantifiable terms.

Carbohydrates

Average per capita consumption of six cereals (wheat, rye, maslin, buckwheat, maize, pulses), potatoes and chestnuts was recorded in the Statistique (Table 11.1). Unfortunately nothing was noted for oats or barley but prefectoral estimates for 1835 provide an introductory view for all eight cereals, albeit a simplified one.[9] This suggested that wheat made up over 80 per cent of the volume of cereals eaten in nine départements in the Paris Basin, Lorraine, Provence and the middle Garonne valley,

Table 11.1: Average Annual per Capita Consumption of Carbohydrates (hl) and Meat (kg)

	Wheat	Maslin	Rye	Maize, etc.*	Potatoes	Pulses	Total
NE	1.99	0.41	0.49	0.11	3.26	0.12	6.38
SE	1.35	0.14	0.93	0.58	2.31	0.09	5.40
NW	1.80	0.50	0.58	0.12	1.94	0.06	5.00
SW	1.64	0.21	0.71	0.44	1.86	0.10	4.96
Continental France	1.72	0.34	0.67	0.27	2.35	0.09	5.44
Corsica	1.67	0.02	0.16	1.15	0.23	0.06	3.29
France Total	1.72	0.33	0.66	0.29	2.34	0.09	5.43

*Maize, buckwheat, chestnuts

	Beef	Veal	Mutton	Lamb	Pigmeat	Goatmeat	Total
NE	7.35	2.25	1.26	0.04	10.71	0.02	21.63
SE	4.93	1.62	3.69	0.45	8.45	0.13	19.27
NW	9.73	2.77	2.52	0.09	7.27	0.03	22.41
SW	3.65	1.85	1.28	0.23	8.45	0.03	15.49
Continental France	6.76	2.19	2.19	0.19	8.66	0.05	20.04
Corsica	3.94	0.11	2.08	0.35	7.83	1.45	15.76
France Total	6.94	2.17	2.19	0.19	8.65	0.06	20.00

but less than 20 per cent in nine départements in the Massif Central and Lower Brittany. Maslin was relatively unimportant but did comprise one-sixth of the volume of cereals eaten in Picardy and the Loire Valley. Inclusion of flour from two or more types of grain was, of course, standard practice in many parts of the French countryside.

In the Massif Central rye made up more than 60 per cent of the volume of cereals eaten, with subsidiary rye-eating areas fringing the Massif to the east and south and being found in eastern Armorica, Champagne and the Landes. Barley consumption was in evidence in the calcareous pays extending from Vendée and Charente-Inférieure to Côte-d'Or and north-eastern France. Armorica was the major focus of buckwheat consumption with another in Limousin, while maize accounted for no less than 66 per cent of the cereals eaten in Basses-Pyrénées and 38 per cent in Hautes-Pyrénées where it was milled to make bread known as *mesture*, formed the base for galettes, and was prepared as gruels. Oats figured in the human diet in a significant way only in Doubs, Aveyron, Finistère and Côtes-du-Nord, being eaten as *grumel* or pottage or as oatcakes. Finally, pulses made up more than one-sixth of the volume of cereals consumed in Somme and Aisne.

Although excluding barley and oats, the Statistique provides detailed information on the statistically determined 'average consumption' at the arrondissement scale of the six remaining grains together with potatoes and chestnuts. These summary figures differed from the real patterns of food consumption of individual families which, as Le Play's monographs displayed so clearly, reflected their wealth, social standing and whether they lived in town or country, as well as their regional location. The largest quantities of wheat entered the statistically average diet in seven groups of arrondissements, where over 2.5 hl were eaten by each inhabitant during the average year (Figure 11.1a). Five groups were in northern France (Ile-de-France, Normandy, Nord, Touraine, Marne plateaux), with the remaining two in Provence and the middle Garonne. Over 3.0 hl were eaten in sixteen arrondissements, rising to approximately 3.5 hl around Saint-Omer (3.42), Toulon (3.47), Provins (3.49) and Auch (3.62). The average residents of many arrondissements containing large cities ate over 2.5 hl of wheat each year but that was not the case for the population of the arrondissements of Lille, Lyons, Nantes or Strasbourg. At the other extreme, the inhabitants of Brittany, the Landes and especially the Massif Central ate very little wheat. Only 0.02 hl was declared to be consumed each year by each inhabitant of Ambert arrondissement where rye was the staple grain, by comparison with 1.69 hl in the neighbouring Limagnes of Clermont-Ferrand.

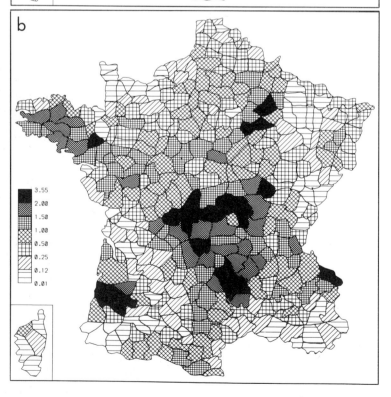

Figure 11.1:
Consumption (a) of
Wheat (b) of Rye
(hl/pc/pa)

a

4.00
2.75
2.25
1.75
1.25
0.75
0.00

b

3.55
2.00
1.50
1.00
0.50
0.25
0.12
0.01

The average Frenchman ate only 0.66 hl of rye each year but four times that amount was consumed in parts of the Massif Central and in Champagne, rising to 3.00 hl around Saint-Flour and 3.55 hl in Arcis arrondissement (Figure 11.1b). As preceding examination of land use and trade would suggest, the pattern was largely the mirror image of that for wheat, with large volumes of rye entering the diet in central and southern Brittany and in a scatter of smaller mauvais pays (such as the Vosges, Sologne, Mauges, Landes, Dombes). In each of these areas relatively little wheat was grown. By contrast, wheat was grown in Dauphiné but was despatched to Lyons so that the local population subsisted on rye. Very small quantities of rye were consumed in Lower Normandy, western Lorraine, Franche-Comté, Provence, Gascony and Saintonge.

Maslin, as such, was not eaten in parts of France as diverse as Paris and the Limousin, however it was quite important in the southern, western and northern parts of the Paris Basin where more than 1.0 hl was consumed by the average inhabitant each year. More than double that amount was eaten in five contiguous arrondissements in Somme and Pas-de-Calais, with maxima around Doullens (2.57) and Abbeville (2.81). As the prefectoral estimates suggested, Armorica and Limousin were the only major foci of buckwheat consumption. Small quantities entered the human diet in Cantal, the Pyrenees and the Sologne and elsewhere buckwheat was used for feeding as grain to poultry and livestock, or even served as a green fodder. A block of three arrondissements in Upper Brittany formed the most important core of buckwheat consumption with an outlier around Mortagne (2.05). Since wheat was exported as a cash crop, buckwheat was, of course, also eaten in coastal Brittany but not in anything like the same quantities as in the interior.

Maize eating was predominantly a feature of the Pays de l'Adour, where over 3.5 hl were consumed each year by each inhabitant in the arrondissements of Orthez (3.59) and Mauléon (3.88). Arrondissements in Landes and Hautes-Pyrénées, together with those in Upper Languedoc, the Garonne valley and Périgord represented the second maize-eating area in the south-west. Not surprisingly, very little maize was eaten in Gers and Haute-Garonne which were important foci of wheat consumption. In the maize growing area of the Saône valley only 0.01 hl per capita was eaten each year. Pulses were eaten in small quantities in all parts of France being mixed with other cereals in bread flour and serving as 'the grain of the poor'.[10] Marche and central Brittany were apparently exceptions to the rule but only in Artois did they achieve any real significance (e.g. Montreuil arrondissement 2.3 hl).

When these data are summated it becomes clear that France

was sharply differentiated in terms of the annual volume of cereals consumed in the 1830s (Figure 11.2). In southern Artois, inner parts of Upper Brittany, and Tarn-et-Garonne more than 5.0 hl were eaten by each resident each year, with 4.0 to 5.0 hl being consumed in parts of the Paris Basin, Touraine and the south-west. At the other extreme, the inhabitants of Ardèche and four arrondissements in Cantal, the Jura and northern Brittany ate less than 1.5 hl of cereals each year and must have obtained their carbohydrates from other crops. The Paris Basin, Upper Brittany, the middle Loire and parts of Aquitaine formed the grain-eating provinces par excellence. Great care should be taken in interpreting Figure 11.2, which must be used at its face value and not to imply a surrogate for total carbohydrate in take. The computations underlying its preparation do not take into consideration the variations in calorific value between the six cereals nor do they include potatoes or chestnuts which were important carbohydrate sources in some areas.

Potatoes were the more significant of the two, being eaten in varying quantities throughout France in the 1830s. More than 3.5 hl were consumed each year by each inhabitant in Finistère, Sarthe, a crescent-shaped area in the Massif Central, Ariège,

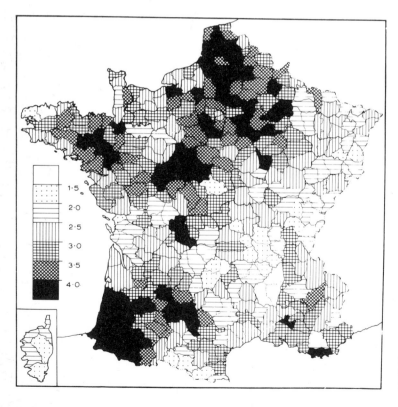

Figure 11.2: Volume of Cereals Consumed (hl/pc/pa)

part of Pas-de-Calais and, most importantly, the north-eastern corner of France. These were the areas where the potato had been adopted early as an escape crop from famine and where, in some cases, earlier rotations had broken down. The highest volumes of consumption were recorded in the northern arrondissements of Bas-Rhin (Weissembourg 10.5, Saverne 10.8), at Saint-Girons (10.9) in Ariège and Le Puy (9.0) in Haute-Loire, where potatoes were eaten as commonly as bread and entered into meals three or four times each day.[11] At the other extreme, potatoes were not of any great importance in three types of region: parts of the south (to which they were unsuited climatically), areas where other staple crops had been adopted recently (e.g. the maize-eating sections of the south-west), and pays, such as parts of Normandy and Beauce, where popular opposition to the potato remained significant in the 1830s.[12] Unlike the widespread growth and consumption of potatoes, the chestnut was very restricted spatially with areas on the south-western margins of the Massif Central being the most important. As was suggested in Chapter 8, information on this crop in Ardèche was underestimated in the Statistique and therefore arrondissements in other parts of central France (Rochechouart 2.21 hl, Tulle 2.39, Brive 3.43 and Saint-Yrieux 4.62) emerge as the major areas of chestnut eating. Either fresh or dried, the crop provided a vital source of food in winter, helping the peasantry of Cantal and Corrèze to survive for four months each year and forming the morning meal in eastern Charente for up to six months.[13]

Volume, per se, is unfortunately a misleading expression of the relative significance of carbohydrate sources and in order to make really meaningful comparisons two types of transformation are necessary. The first converts the volume of each commodity consumed into its appropriate weight and represents an intermediate step in generating calorific values (Table 11.2). Conversion from volume to weight has been achieved by using the national multipliers for each crop listed in the agricultural census of 1852 which are the closest available in time to the Statistique of the July Monarchy. Two problems are raised by adopting this procedure: first, the 1852 multipliers were probably fractionally higher than would have been the case in the 1830s; and second, the multipliers for each crop would vary from region to region. Undoubtedly a more meaningful transformation is the conversion of weights of commodities into calorific values. This process has the effect of shrinking the importance of potatoes and (to a lesser extent) of chestnuts in the pattern of carbohydrate consumption. For whereas the conversion factors for cereals ranged from 346 calories for every 100 g of wheat to 355 calories for every 100 g of maize, the factor was 250 for

Table 11.2 Conversion Factors for Carbohydrates, Beverages and Meat

Carbohydrates[1]	Average weight (kg) per hl (1852)	Average calories per 100 g
Wheat	73.7	346
Maslin	69.4	350
Rye	68.2	350
Barley	58.9	339
Oats	43.3	380
Buckwheat	55.7	348
Maize	70.3	355
Pulses	73.7	340
Potatoes	68.3	75
Chestnuts	68.3	250

Beverages[2]	Average calories per l
Wine	86
Beer	40
Cider	40
Eau de Vie	220

Meat[3]	Average calories per kg (moderately fat, whole carcase)
Beef	262
Veal	184
Mutton	249
Lamb	249
Pigmeat	420
Goatmeat	145

Sources: 1. *Statistique de la France, 1852*, cited in Ministère de l'Agriculture, *Statistique agricole de la France, 1882* (Nancy, 1887); B.S. Platt, 'Tables of representative values of foods commonly used in tropical countries', *Medical Research Council Special Report Series*, vol. 302 (1962); Food and Agricultural Organisation, *Food Composition Tables* (Rome, 1954); Anon., 'The economic relationships between grains and rice', *Food and Agricultural Organisation Commodity Bulletin Series*, vol. 39 (1965).
2. S. Thomas and M. Corden, *Tables of Composition of Australian Foods* (Canberra, 1970).
3. Platt, 'Tables of representative values'.

chestnuts but only 75 for potatoes. Since the Statistique provided figures on gross consumption an allowance has to be made for the proportion of grain lost in milling. This would, of course, vary according to the sophistication of the milling technique but an average reduction of ten per cent has been applied throughout and the same proportional reduction will be used in the conversion of weights of meat into quantities of calories consumed.[15]

These data on the calorific value of carbohydrate consumption in each arrondissement have been used to prepare a combination map following the modified Weaver technique outlined in Chapter 5. When the pattern of carbohydrate consumption conformed to either a one- or two-element model these features are indicated by shading. If a three-element model was appropriate the third has been identified by a literal symbol. A number of carbohydrate 'provinces' may be recognised on the basis of the relative calorific importance of the various crops consumed (Figure 11.3a).

Wheat emerged as the leading source of calories derived from carbohydrates across the greater part of the country, with 'wheat' in isolation dominating six areas: a large zone extending through Provence, Lower Languedoc, the middle Garonne and the pays charentais to Lower Poitou; Normandy; Nord; Brie plus the environs of Paris; Haute-Marne; and the Bresse. In terms of calories, wheat was the leading carbohydrate source in most other parts of France with various two- or three-element models describing a large number of combinations. Thus 'wheat and maize' typified parts of eastern Aquitaine and the Dordogne, and 'wheat and rye' Champagne, Dauphiné and a wide stretch of middle France from Dijon to the mouth of the Loire. The rye province covered much of the Massif Central, the Landes, parts of Champagne, southern and central Brittany, the fringes of Armorica and the high Alps.

The remaining provinces were tightly regionalised, with maize being the leading element in only ten arrondissements in the Pays de l'Adour plus Sarlat in Périgord, buckwheat dominating seven arrondissements in Upper Brittany and Lower Normandy, and maslin appearing in leading position in parts of Picardy and two sections of the Paris Basin. Chestnuts formed the leading source of calorie intake in three arrondissements of Limousin and potatoes provided the leading source of carbohydrate in sixteen arrondissements, seven of which were in the Vosges, one in Alsace, three in Lower Brittany, two in Ariège, two in the eastern Massif Central and the remaining one in Anjou.

Figure 11.3a has the disadvantage of only being able to display models with up to three elements. To reduce this problem somewhat Figure 11.3b has been prepared which identifies the number of elements generated for each arrondissement before drawing up Figure 11.3a. Polyculture was most significant on the western margins of the Massif Central with no fewer than five elements providing the best fit in the arrondissements of Riberac, Sarlat and Limoges, with a four-element model describing a surrounding cluster of arrondissements in Corrèze, Dordogne, Haute-Vienne and Lot. At the other extreme, the single element 'wheat' areas, the two 'maize' arrondissements of the Pays de l'Adour and the seven 'rye' arrondissements of the Massif Central emerge most clearly.

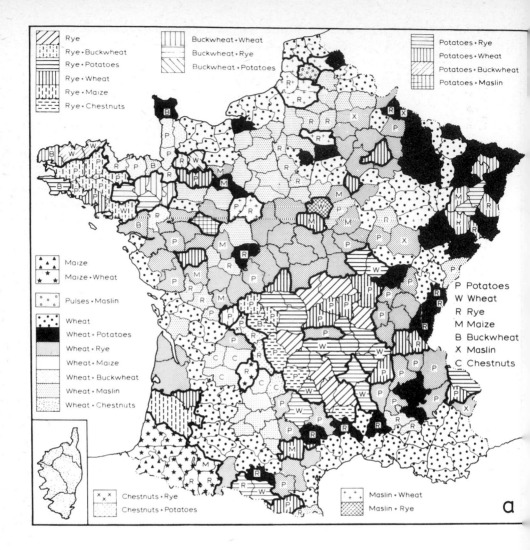

Legend entries:

Rye
Rye + Buckwheat
Rye + Potatoes
Rye + Wheat
Rye + Maize
Rye + Chestnuts

Buckwheat + Wheat
Buckwheat + Rye
Buckwheat + Potatoes

Potatoes + Rye
Potatoes + Wheat
Potatoes + Buckwheat
Potatoes + Maslin

Maize
Maize + Wheat
Pulses + Maslin
Wheat
Wheat + Potatoes
Wheat + Rye
Wheat + Maize
Wheat + Buckwheat
Wheat + Maslin
Wheat + Chestnuts

P Potatoes
W Wheat
R Rye
M Maize
B Buckwheat
X Maslin
C Chestnuts

Chestnuts + Rye
Chestnuts + Potatoes

Maslin + Wheat
Maslin + Rye

a

Beverages

Alcoholic beverages formed the second component of dietary consumption described in the Statistique. Greatest information was given for wine, which was drunk by some but not all sections of society virtually nationwide. Peasants with a vine or two consumed at least part of what they produced even though it may have been watered down *petit vin* as in the Limagnes, whereas water was the normal drink for the mountainfolk of the Auvergne with only a little wine of the poorest kind.[16] In spite of the fact that vines were grown more widely and wine drunk more extensively than at the start of the century Blanqui might still assert that wine drinking was still unknown to three-fifths of the peasantry in 1850.[17] The consumption of beer, cider and eau

Figure 11.3: (a) Carbohydrate Provinces (b) Number of Elements

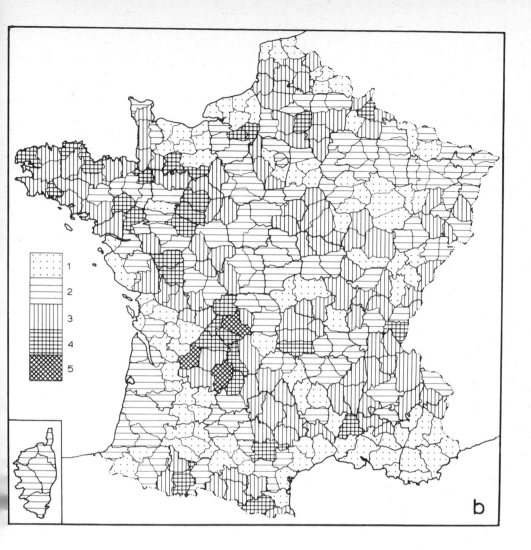

b

de vie (including Calvados) had distinctive regional patterns. Beer drinking was a phenomenon of the northern and north-eastern départements so that, with the exception of Nord (1.54 hl), Ardennes (0.95) and Pas-de-Calais (0.39), the volume consumed by an average inhabitant during the course of a year was slight. Only 0.30 hl of cider was drunk per capita each year for the nation as a whole, but in fact most of this occurred in Normandy, with over 2.0 hl being consumed per annum by the average inhabitant of Manche (2.77), Calvados (2.27), Orne (2.12) and Eure (2.09). A wider zone of cider country extended into Brittany, Upper Normandy and Picardy. Eau de vie was essentially a northern drink, with consumption of Calvados brandy accentuating the role of Normandy and alcoholism being the 'dominant vice' in Perche and neighbouring pays.[18] Unfortunately no distinction was drawn between types or qualities of wine in the Statistique, but in terms of volume, the residents of Hérault (2.50 hl), Loir-et-Cher (2.44), Gers (2.32) and Yonne (2.26) headed the league. No figures for wine consumption were given for départements in northern and north-western France that were beyond the limit of vinegrowing, where beer or cider were the popular beverages. The average Parisian (1.25 hl), Lyonnais (1.27) or Marseillais (1.21) drank slightly less than twice the national average (0.70) but consumption by these urban dwellers was small by comparison with the declared intake of the inhabitants of the arrondissements of Béziers (3.49), Bar-sur-Seine (3.87) and Blois (4.37). As well as providing relatively cheap local supplies, the vineyard areas also witnessed the use of wine as a source of payment in kind for agricultural workers. This contributed to the very high rates of consumption in the Midi and some other areas. In any case, low quality wine ensured a supply of cheap calories every time bread was short. Not surprisingly, the fine wines of Burgundy and Bordeaux were not consumed in large quantities by the local population. At the other end of the wine-drinking spectrum, less than 0.05 hl was drunk each year by the average inhabitant of Velay, the *planèze* of Saint-Flour and some arrondissements along the northern margin of vine cultivation.

Meat

Consumption of no fewer than six types of meat was recorded in the Statistique. Beef was essentially a feature of diets in the population of northern parts of the country and of urban areas (Figure 11.4a). The average Frenchman ate 6.94 kg each year but in Seine the figure was over four times the national average (30.0 kg). Beef eating was impressive in other arrondissements that contained large cities, for example Versailles (24.3), Strasbourg (21.0), Rouen (18.9), Lyons (18.7), Brest (14.2), Bordeaux (14.1), Toulon (12.9), Avignon (11.7) and Nîmes (10.4). All other arrondissements in which beef consumption exceeded 9.0 kg per year were located in the Paris Basin and in the Lower Seine. By contrast with these areas, the inhabitants of many parts of southern France ate less than 1.0 kg of beef each year, with consumption in southern sections of the Massif Central falling to half that amount or less. A fascinating local contrast existed between the quantity of beef eaten by the average Lyonnais (18.7) and that consumed by his Stéphanois neighbour (9.9) whose diet was an urbanised reflection of the peasant regime of the eastern Massif Central. None the less, the message is clear. During the July Monarchy beef was eaten in relatively large quantities by inhabitants of the Ile-de-France, in Nord and Alsace but in virtually no other parts of the country.

Figure 11.4:
Consumption (a) of
Beef and (b) of Mutton
(kg/pc/pa)

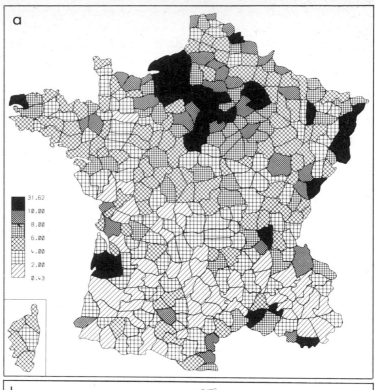

a

	31.62
	10.00
	8.00
	6.00
	4.00
	2.00
	0.43

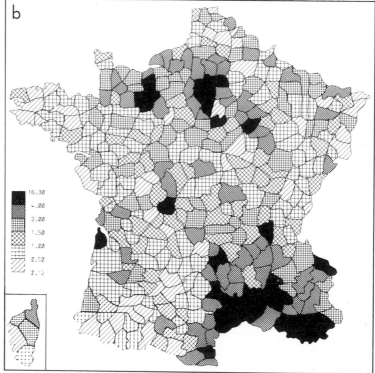

b

	16.30
	4.00
	2.00
	1.50
	1.00
	0.52
	2.12

Only 2.1 kg of veal were eaten each year by the average Frenchman and, not surprisingly, the pattern of consumption was very similar to that for beef (Bordeaux 6.3, Lyons 5.9, Seine département 5.9), with the addition of a number of stock-raising localities where surplus calves were slaughtered and entered the local diet in that way. This process accounts for the large quantities consumed in Cantal (Murat 5.2, Aurillac 4.8), in Upper Brittany (Rennes 5.5) and Lower Normandy (Lisieux 9.6). Veal was virtually unknown in the Mediterranean south, with consumption falling to less than 0.02 kg in Upper Provence.

It was, of course, mutton that formed the main source of meat in the south, with residents of Montpellier (16.30 kg) and Marseilles (14.21) consuming six or seven times the national average (2.19) and figures remaining high throughout the countryside of the Midi (Figure 11.4b). The Ile-de-France and Oise formed the second focus of mutton eating, with as much as 9.34 kg being eaten each year by the average inhabitant of Seine and 3.0 to 5.0 kg being consumed by his rural neighbour. Like mutton, lamb was eaten in large quantities in the Midi, with surplus lambs born in the Camargue being consumed in surrounding arrondissements such as Arles (2.06), Avignon (2.36) and Nîmes (3.61). These figures appear truly enormous when set against the national average of 0.19 kg and the statistics for sections of the Massif Central, Artois and southern parts of the Paris Basin where lamb eating was virtually non-existent. The arrondissements of Corsica (1.0 to 1.89 kg) and Grasse (2.28) in Var were the only parts of France where goat meat was eaten in any quantity.

By contrast with the five preceding types of meat, pig meat was eaten throughout the country and formed the only kind of meat that was likely to appear on the tables of the poor. The average Frenchman consumed no less than 8.65 kg each year and in north-eastern France more than double that quantity was eaten, with the largest amounts being consumed around Metz (20.2), Châlons-sur-Marne (21.2) and Briey (23.1) (Figure 11.5a). The population of the southern margins of the Massif Central and Gironde ate between 13.0 and 16.0 kg each year but beyond these areas the eating of pig meat was essentially a feature of the countryside. Many arrondissements containing large cities recorded only average or below average levels of consumption (e.g. Lyons 8.9, Seine département 8.0, Lille 4.9, Marseilles 2.8). Relatively little pig meat was eaten in sheep-rearing regions such as Provence and Berry or in the 'progressive' agricultural environments of Normandy, Brie and Nord.

The average Frenchman ate 20.0 kg of meat each year in the 1830s which Haumont believed to be only one-third of what the

Figure 11.5:
Consumption (a) of
Pigmeat and (b) of All
Types of Meat
(kg/pc/pa)

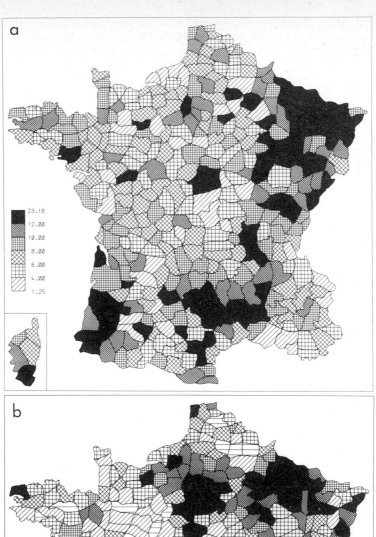

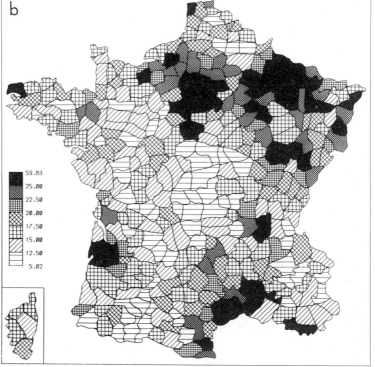

average Englishman consumed.[19] The average inhabitant of Seine ate 55.6 kg of meat and residents of Saint-Denis consumed even more (74.0 kg) (Figure 11.5b). Around Lectoure in Gers the quantity fell to 5.8 kg, one quarter of the national average. The inhabitants of twenty-three arrondissements ate over 30.0 kg of meat and it was in Versailles (47.5), Lyons (42.3), Bordeaux (42.0), Strasbourg (41.0) and Châlons-sur-Marne (40.0), as well as in the capital, that the meat eaters resided. Their country cousins in the Morvan and in parts of the Massif Central and Aquitaine got by on less than 10.0 kg each year. As during the ancien régime, 'for the really poor, Lent lasted for the whole year round'.[20]

As with the sources of carbohydrate there are of course important variations in the food value of different types of meat. Pig meat has a high calorific co-efficient, veal and goat meat have low multipliers, whilst beef and mutton occupy intermediate positions (Table 11.2). In fourteen arrondissements more than half of the calories derived from meat came from beef, of which eight were in Seine and Seine-Inférieure. In the Pays de Caux the proportion exceeded 60 per cent but it fell to under ten per cent in predominantly rural arrondissements in the southern third of France. Veal was relatively insignificant as a source of calories, being almost entirely absent from the south. Unlike beef, its role was most important in pastoral areas rather than in large cities and it contributed over 15 per cent of meat-derived calories in Cantal, Gers, the Bresse, the Pyrenees and Marche. Mutton provided over ten per cent of meat-derived calories in parts of the Ile-de-France, Normandy and Perche but especially in the Midi. Its importance increased with proximity to the Mediterranean coast, contributing two-fifths of the calorie intake from meat around Béziers and Montpellier and a half at Marseilles and Aix. Yet it was the northern arrondissement of Lisieux (55 per cent) that came in leading position. Goat meat was insignificant except around Grasse (8 per cent) and in southern Corsica (Sartène and Ajaccio 5 per cent). Pig meat provided more than 50 per cent of meat-derived calories across two-thirds of France, rising to three-quarters in north-eastern parts of the country, the Alps, southern parts of the Massif Central, Vendée and sections of Aquitaine and Armorica, and even reaching 90 per cent in the neighbouring arrondissements of Rodez, Villefranche and Figéac and around Oléron in the Pyrenees. The calorific contribution of each type of meat to the total number of calories derived from all types of meat may be calculated in the same way as that described for land-use combinations. Beef was the leading source of meat-derived calories in only 42 arrondissements, with 40 of these being characterised by 'beef and pig meat'. The most extensive clusters of beef-eating areas were in Normandy, the Ile-de-France and Nord, with isolated examples in Berry, around Lyons, Toulouse, La Tour and Dôle. Only the contrasting pastoral environments of the Camargue (Arles) and the upper Durance (Briançon) displayed the 'beef and mutton' combination. Mutton (plus lamb) was the leading calorie source in only five arrondissements, with 'mutton and pig meat' in Aix, Lodève and Montpellier arrondissements and 'mutton and beef' around Lisieux and Marseilles. Pig meat provided the greatest source of meat-derived calories in every other part of France, with intake approximating most closely to the one-element 'pig meat' model in a great U-shaped stretch of territory from Cotentin to Aquitaine, through the southern Massif Central and then northwards to Lorraine. Most of the remaining parts of the country displayed two-element patterns with 'pig meat and beef' in most of the Paris Basin, Armorica, the northern Massif Central and Alsace; 'pig meat and mutton' in the Mediterranean south; and 'pig meat and veal' in Guéret (Creuse) and Lectoure (Gers).

The average quantity of calories derived from the six types of meat listed in the Statistique amounted to 58,000 per capita/per annum. In four regions more than 72,000 calories were derived from meat in the average annual diet (Figure 11.6). Lorraine plus surrounding parts of north-eastern France formed the most extensive of these, being followed by the Ile-de-France, the lower Rhône and Lyonnais. Virtually all other areas with such large quantities of calories derived from meat either contained major cities (e.g. Bordeaux, Rouen, Strasbourg, Toulon) or involved distinctive pastoral environments (e.g. Aubrac or Boulonnais). In fact, in eight arrondissements the average inhabitant consumed over 108,000 meat-derived calories each year, with the record being held by Paris and its suburbs (132,000 calories). In stark contrast the average meat consumption in a dozen arrondissements of central, south-western and Alpine France amounted to less than 27,000 calories, falling to under 18,000 around Lectoure (16,240), Bourgneuf (17,760), and Barcellonnette (15,780) where the peasantry endured 'the greatest suffering'.[21] Meat-eating France meant the capital, the north-east and Languedoc.

In complete contrast, average annual diets in Picardy, Touraine, Limousin, the middle Garonne and the extreme south-west involved intakes of over 1,000,000 calories derived from carbohydrates, which rose as high as 1,360,800 around Bellac and 1,564,000 around Saint-Yrieux. The national average was 792,000 per capita/per annum. Average diets in only three arrondissements in the Ile-de-France exceeded 1,000,000 calories derived from carbohydrates but in many arrondissements of that region the 900,000 calories mark was surpassed.

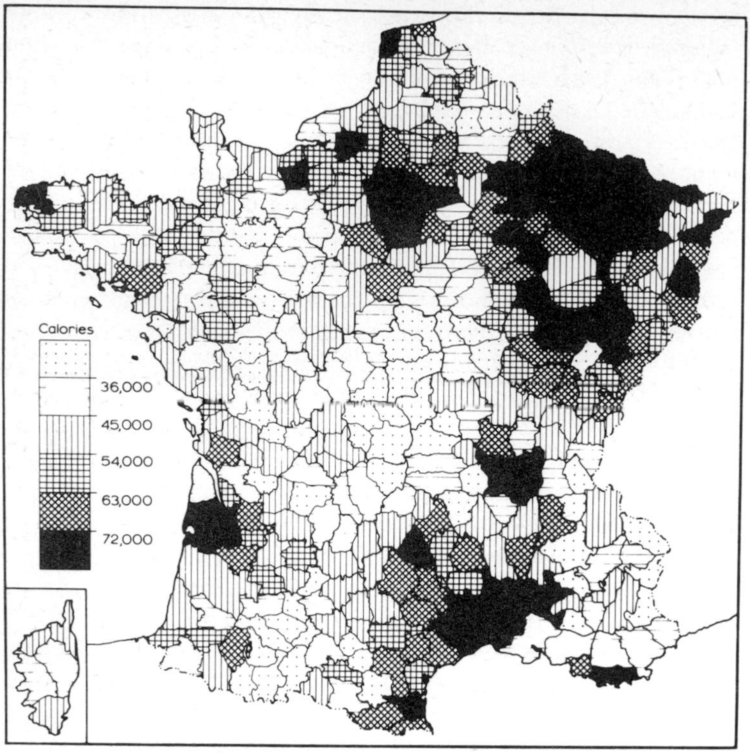

Figure 11.6:
Calories Derived from
Meat (pc/pa)

Champagne, Franche-Comté, the pays charentais and eastern
and southern parts of the Massif Central came at the opposite
end of the carbohydrate scale, with less than 735,000
carbohydrate-derived calories each year (or 210 kg of grain/per
capita/per annum) which, according to Clark and Haswell, 'can
be called the subsistence minimum'. [22]

Want and Plenty

According to the evidence of the Statistique, diets were overwhelmingly vegetarian throughout France, even in Paris, in the
1830s. In only eleven arrondissements (in Champagne and
Franche-Comté and around Saint-Etienne) did the six types of
meat listed contribute the equivalent of more than 20 per cent of
the calories derived from carbohydrates. In the immediate environs of Paris, western Lorraine, Alsace, Burgundy, Lyons,
Ardèche, Gard and Hérault meat provided the equivalent of 10
to 20 per cent, but fell to 2.5 to 5.0 per cent in Artois and large
stretches of central, south-western and Alpine France. In four
arrondissements it even fell below 2.5 per cent. In spite of the

208

conventional wisdom that meat eating was more pronounced in urban areas than in the countryside, most arrondissements that contained large cities were characterised by only moderate ratios. The reason for such a situation resided with the social diversity of these urban areas, with the rich diets of the bourgeoisie being more than counter-balanced by the meagre regime of the urban proletariat and the residents of the peri-urban countryside. Olwen Hufton's assertion that '95 per cent and upwards' of the diet of the poor in eighteenth-century France was derived from cereals would also seem to hold true for sizeable sections of French society during the July Monarchy. [23] Such an emphasis on cereals implied that 'even if one ate one's fill the supply of protein and vitamins was grossly inadequate'. As in earlier times 'rickets, scurvy and allied deficiency diseases were the normal lot of families whose stomachs were full and hence were unaware of hunger pangs'. More educated contempories were well aware of the dangers of 'insufficient, poor and exclusively cereal-derived food'. [24]

There was undoubtedly also much truth in Lullin de Châteauvieux's claim that the French ate more cereal but less vegetables, meat or milk than the inhabitants of any other European country. [25] A precise answer may not be given for although fish, dairy products, game and fruit entered into French diets in the 1830s their role may not be quantified. Populations in coastal, marshy or riverine areas used fish as a source of protein and along the Mediterranean coast the resources of the sea compensated to some extent for the lack of pasture which prevented large numbers of livestock being raised. Dairy products, especially in the form of skimmed milk, were consumed in cereal-based soups and gruels throughout the country but they were particularly significant in pastoral areas where cheese made from the milk of cows, ewes or, in the case of Corsica, goats was also eaten.

Eggs were consumed in some parts of the countryside but were often sold to towns or reserved for invalids. Game was hunted, if not poached, throughout France and added valuable supplies of protein to the diets of the peasantry in areas of marsh, moor and wood and to the food consumption of the inhabitants of small towns where it was eaten instead of butcher's meat. Indeed game was even supplied on a commercial basis to the capital from a dozen surrounding départements. [26] Fruit and wild berries could offer yet more variety according to the particular environmental resources of the locality. Unfortunately it is impossible to quantify their consumption on a regional basis, even though some estimates have been made for France as a whole. [27] Similarly, little is known of the pattern of poultry consumption, although the *basse-cour* must have made a useful contribution.

All that may be done is to produce an indication of the total annual intake of calories derived from carbohydrates, six types of meat and alcohol and then stress that the picture is unavoidably incomplete. On the basis of this evidence the average French citizen consumed 857,000 calories per annum, whilst his counterpart in Limousin, the Pays de l'Adour and the middle Garonne was consuming more than 1,170,000 calories each year (Figure 11.7). In fourteen arrondissements the total rose in excess of 1,260,000, with absolute maxima being around Brive (1,443,400), Saint-Pol (1,484,000), Arras (1,496,900) and Saint-Yrieux (1,615,000). Such levels were in excess of the range between 1,277,500 and 1,460,000 calories (or 3,500 to 4,000 per day) which was what Braudel described as necessary for 'a man [of] today [. . . who] belongs to a rich country and a privileged class'.[28] In harsh contrast were the 'hungry' pays, with annual calorie intakes of less than 720,000 applying in Champagne, Franche-Comté, the Morvan, Creuse, Cantal, Forez, Ardèche and Léon. Diets in the central provinces were unquestionably the most deficient and unhealthy. Absolute minima related to the arrondissements of Dôle (321,080), Pontarlier (329,660) and

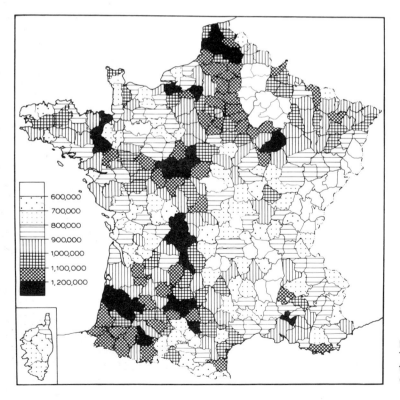

Figure 11.7:
Total Intake of Calories from Specified Sources (pc/pa)

210

Saint-Claude (350,000) which fell far below the figure of 822,000 calories (or 2,800 per day) which Toutain quoted as the 'recognised normal level of the average man'.[29] As far as carbohydrates, meat and alcohol were concerned, the diet of average residents in parts of Artois and Limousin appeared to contain no less than four times as many calories as those of their counterparts in the Jura.

It is only reasonable to add some qualification to the information displayed on Figure 11.7. First, the low calorie intake of Léon, as with all other coastal areas, would be enhanced by fish in the diet; and second, in Franche-Comté and Cantal one might anticipate some consumption of barley and also important inputs from dairy products since such areas had high densities of cattle. Additions of this kind were less likely in Champagne, Upper Poitou, the southernmost arrondissements of the Massif Central, central Corsica and the southern Alps. Here was 'hungry' France where communications were poor, the need for self-sufficiency formed the harshest fact of life, and traditional sources of food remained relatively unchanged until well after mid-century. These were the areas that were most liable to suffer when local food supplies failed, as they had done many times in the past and were to do again with disastrous consequences in 1846-7. 'De la famine, délivrez-nous, Seigneur!' remained the fervent prayer of a population whose 'principal recreation', to quote Le Play, 'consisted of adding a little meat on *jours de grande fête* and perhaps on Sunday to the normal regime of bread, water, butter and potatoes'.[30]

Parisians, on average, ate well during the July Monarchy and, as Husson and Aron (more than a century later) have shown, some ate extremely well and probably better than the citizens of any other European city.[31] By contrast, 'the diet of most peasants is inferior to those in most other nations. No meat, white bread or wine; only black bread, potatoes and water; this is how a third of the French people eats'.[32] Such was the observation of Léonce de Lavergne twenty years after the Statistique. There were, of course, hungry and impoverished sections of society throughout the country whose plight has been subsumed so easily in statistics for the mythical average inhabitant.[33] The *petits ménagers* who live around Douai provide a salutary corrective.[34] They subsisted on potatoes and a little bread but never ate meat, even though they lived in Nord département which was characterised by intensive, commercial farming and food supplies that were 'good and healthy' and included white bread, butter, cheese, potatoes, beef and pig meat.[35] The petits ménagers of Douai were not alone, for their counterparts were the small peasants and landless labourers of both town and country in every pays of France.

Notes

1. F. Braudel, *Capitalism and Material Life, 1400-1800* (London, 1973), p. 89.

2. M. de Tocqueville, 'Des principales questions agricoles: bestiaux, vins, laines', *JAP* 2nd series, vol. 1 (1843-4), p. 100.

3. J. Vidalenc, *Le Peuple des Campagnes: la société française de 1815 à 1848* (Paris, 1970), p. 65.

4. M. Brieude, 'Observations économiques et politiques sur la chaîne des montagnes ci-devant appelées d'Auvergne', *AS*, vol. 3 (1802), p. 380.

5. M. Michaux, 'Le département du Puy-de-Dôme en 1841-42 d'après les mémoires et reconnaissances militaires', *ACSS* (1963), p. 834.

6. J.J. Hemardinquier, *Pour une Histoire de l'alimentation* (Paris, 1970); A range of articles appearing in *AESC* vol. 16 (1961) and vol. 30 (1975) and in other periodicals is cited in H.D. Clout, *Agriculture in France on the Eve of the Railway Age* (Ph.D. thesis, University of London, 1979).

7. F. Le Play, *Les Ouvriers européens* (6 vols., Paris, 1879).

8. B. Bennasar and J. Goy, 'Contribution à l'histoire de la consommation alimentaire du XIVe au XIXe siècle', *AESC*, vol. 30 (1975), pp. 402-29.

9. Ministère des Travaux Publics, de l'Agriculture et du Commerce, *Archives statistiques* (Paris, 1837).

10. Picot de la Peyrouse, *The Agriculture of a District in the South of France* (London, 1819), p. 49.

11. M. de Sainte-Colombe, 'Notice sur l'instruction publique, l'agriculture et l'industrie dans l'arrondissement d'Yssingeaux', *Annales de la Société d'Agriculture, Sciences, Arts et Commerce du Puy*, vol. 3 (1838), p. 106.

12. G. Désert, 'La culture de la pomme de terre dans le Calvados au XIXe siècle', *AN*, vol. 5 (1955), pp. 216-70; J. Doublet de Boisthibault, *La France: description géographique, statistique et topographique: Eure-et-Loir* (Paris, 1836), p. 142.

13. Brieude, 'Observations Auvergne', p. 393; A. Firmigier, 'Essai de statistique du département de la Corrèze', *AS*, vol. 4 (1802), p. 184; J.P. Quénot, *Statistique du département de la Charente* (Paris, 1818), p. 338; E. Appolis, 'L'alimentation des classes pauvres dans un diocèse languedocien au XVIIIe siècle', *ACSS*, (1968), p. 56.

14. Cited in Ministère de l'Agriculture, *Statistique agricole de la France, 1882* (Nancy, 1887).

15. C. Clark and M. Haswell, *The Economics of Subsistence Agriculture*, 4th edn, (London, 1970), p. 57.

16. D. Bertrand, 'Propositions générales sur la topographie médicale du département du Puy-de-Dôme', *ASLIA*, vol. 21 (1848), p. 436.

17. Cited in E. Weber, *Peasants into Frenchmen: the modernization of rural France, 1870-1914* (London, 1977), p. 144.

18. Vidalenc, *Le Peuple*, p. 172; J. Vidalenc, 'Les habitants de l'Orne au début de la monarchie de Juillet', *AN*, vol. 15 (1965), p. 583.

19. M. Haumont, 'Production animale de la France et de l'Angleterre', *JAP*, vol. 4 (1846-7), p. 16.

20. R. Mandrou, *Introduction to Modern France 1500-1640: an essay in historical psychology* (London, 1975), p. 17; P. Souty, 'Le Carême dans les établissements réligieux tourangeaux et blésois au XVIIIe siècle', *ACCS* (1968), pp. 115-24.

21. B. Chaix, *Préoccupations statistiques, géographiques, pittoresques et synoptiques du département des Hautes-Alpes* (Grenoble, 1845), p. 800.

22. Clark and Haswell, *Subsistence Agriculture*, p. 59.

23. O. Hufton, *The Poor of Eighteenth Century France* (Oxford, 1974), p. 44, p. 45.

24. V. Nivet and H. Aguilhon, 'Notice sur l'épidémie cholérique', *ASLIA*, vol. 23 (1850), p. 389.

25. F. Lullin de Châteauvieux, *Voyages agronomiques en France* (Paris, 1843), p. 23.

26. G. Léonce de Lavergne, *L'Agriculture et la population* (Paris, 1857), p. 391.

27. J.C. Toutain, 'La consommation alimentaire en France de 1789 à 1964', *Cahiers de l'I.S.E.A.: Economies et Sociétés*, vol. 5, no. 11 (1971), pp. 1909-2049.

28. Braudel, *Capitalism*, p. 87.

29. Toutain, 'La consommation alimentaire', p. 2031.

30. E. Juillard (ed.), *Histoire de la France rurale: apogée et crise de la civilisation paysanne 1789-1914* (Paris, 1976), vol. 3, p. 134; Le Play, *Les Ouvriers*, vol. 6, p. 201.

31. A. Husson, *Les Consommations de Paris* (Paris, 1856); J.P. Aron, *Le Mangeur au XIXe siècle* (Paris, 1973).

32. Léonce de Lavergne, *L'Agriculture*, p. 298.

33. R. Philippe, 'Une opération pilote: l'étude du ravitaillement de Paris au temps de Lavoisier', *AESC*, vol. 16 (1961), pp. 564-7.

34. Inspecteurs de l'Agriculture, *Nord*, p. 30.

35. N. Grar, 'Nourriture', *La Flandre Agricole et Manufacturière*, vol. 1 (1834-5), p. 245; L. Trenard, 'L'alimentation en Flandre française au XVIIIe siècle', *ACSS* (1968), pp. 59-113.

The Value of Production

The literary record of French farming in the past century is composed of a wealth of detailed description focusing on single départements and individual pays. Few authors cast their gaze more widely and consequently definitions of major agricultural regions were rare indeed. When such an approach was attempted départements were usually rearranged into groups but little explanation was given of the logic behind the operation.[1] Adolphe Blanqui adopted a different method, taking major rivers as dividing lines between supposedly different country-sides, while Lullin de Châteauvieux chose to ignore both administrative boundaries and watercourses.[2] Instead, he paid considerable attention to the topographic and botanic composition of France as well as to spatial variations in agricultural conditions, incorporating a lengthy justification for the eight agricultural regions that he defined.

The labours of mayors, sub-prefects and prefects throughout Louis-Philippe's kingdom and the calculations by the clerks of the Ministry of Commerce and Public Works have bequeathed an ample but tantalising legacy which has been mapped to depict individual patterns of land use, production, yields, prices, surpluses and shortages for a wide range of commodities. All this information on crops, livestock and firewood is not amenable to synthesis in its original state and requires transformation to a standard form of expression such as the combined value of output expressed in francs/hectare. Such an operation has been undertaken to provide an alternative view of the structure of French farming in the July Monarchy to those that were assembled by contemporary or near-contemporary writers. It also introduces a wider perspective than that offered by Michel Morineau who concentrated on cereal productivity to the exclusion of other aspects of farming.[3]

Information in the Statistique may be classified into three groups according to its suitability for manipulation. The first group includes natural grass, artificial meadows, fallows, firewood, the 'special crops' and all forms of livestock for which the value of output (or, in the case of livestock, the total value) is given for each arrondissement. Pâtis, communaux, landes and bruyères form the second group. In this case no information on value (or, indeed, area) was given by arrondissement. Instead the average value by département has been related to the figure for 'other land uses' in each arrondissement that was presented in Chapter 5. The third group comprises each of the arable crops for which the value of net yield/ha has been taken as the basis for calculation. The national value of all forms of production including alcohol amounted to 6,468,719,000 F, giving an average value of 126 F/ha. The contribution of each crop and each form of livestock to the national total expressed in the Statistique has been summarised into a number of categories which are depicted in Table 12.1.

Table 12.1: Value of Production

		Francs	%
Firewood		206,600,525	3.19
Vines	wine	419,029,152 ⎫	7.39
	eau de vie	59,059,150 ⎬	
Fodders	natural grass	462,598,243 ⎫	
	artificial meadows	203,765,169 ⎪	12.99
	fallows	92,285,902 ⎪	
	rough grazing	82,064,046 ⎭	
All other crops		3,001,495,000	46.40
All livestock		1,941,824,000	30.01
Total Value		6,468,721,000	
Pastoral realm*		3,084,549,830	47.67

*Fodders, all livestock, and oats

Firewood accounted for a mere 3.19 per cent of the specified total value of production but its role was greatly enhanced in the north-eastern quarter of France where a block of arrondissements between Bourges, Belfort and Rocroi owed over eight per cent of the value of their total production to firewood and more than twice that proportion characterised areas in the Ardennes and around Châtillon (Côte-d'Or), Saint-Claude (Jura) and Sartène (Corse). In the arrondissements of the Landes, Brignoles (Var), Saint-Marc (Isère), Rambouillet (Seine-et-Oise) and Coulommiers (Seine-et-Marne) firewood also accounted for more than eight per cent of the total value of production but contributed less than one per cent in much of Armorica, Vendée, the pays charentais, the southern margins of the Massif Central, the Rhône delta, Ain and Flanders.

Hay contributed 7.15 per cent of the total value of production and more than one-eighth in a great U-shaped stretch of arrondissements extending along the western, southern and eastern margins of the Massif Central, thence into the meadows of the Dombes and the Bresse. High mountain pastures in the Alps, central Pyrenees, Jura and Vosges contributed a similar proportion, as did the marshy grasslands along the west coast and in the vales of Lower Normandy. Around Pont-l'Evêque (25.5 per cent), Lisieux (25.6 per cent), Castellane (27.3 per cent) and Rémicourt (29.0 per cent) hay contributed the highest propor-

tions to the total value of production, whilst in the arrondissements of Arras, Brignoles, Narbonne, Pithiviers, Uzès and
Villefranche the fraction fell to less than one per cent.

The products of the vine, in the form of wine and eau de vie,
represented 7.39 per cent of the value of national output and in
eight areas they exceeded 20 per cent of the total. The largest of
these involved five arrondissements in Bordelais, with very high
proportions being recorded around Blaye (47.7 per cent),
Bordeaux (45.4 per cent) and Libourne (40.9 per cent). Smaller
concentrations involved the plain of Languedoc, western Var,
the pays charentais, Beaujolais, Mâconnais, parts of Aube, and
the environs of Colmar and Orléans. More than 10 per cent of
the value of production was wine-based over the greater part of
Aquitaine, the upper Saône, sections of the Loire valley, Yonne,
Champagne, the Limagnes and the arrondissements of Toul,
Nancy and Versailles. The contribution of the vine was slight in
major upland areas, in the Bresse plain and near the limit of
cultivation from the Gulf of Morbihan to the Ardennes.

Livestock of all kinds made up 30.01 per cent of the value of
production throughout the country with virtually all areas falling into the range between 30 and 39 per cent. Important cattle-
raising areas, such as Cantal, Manche, Maine and the Morvan,
as well as dairying pays in Normandy and the milksheds of
Seine, exceeded 40 per cent, with values just over 50 per cent
around Boussac, Céret, Prades and Pont-l'Evêque and even
higher percentages for the urbanised arrondissements of
Marseilles (55.2 per cent) and Paris (94.4 per cent). For Seine as a
whole the proportion was 58.9 per cent. By contrast, in
Languedoc, Gironde and the lower Rhône animals contributed
less than 20 per cent of the value of production across extensive
areas. This was in part because of the financial importance of the
vine or other high-value crops like mulberries but also because
of more fundamental fodder deficiencies in much of the south.

Arable farming plus all special crops except the vine completed the picture by contributing 46.40 per cent of the national
value of production as specified in the Statistique. In very few
areas did this category exceed 70 per cent and in such cases its
role was enhanced either by very high cereal yields, for example
in five arrondissements of Artois and Cambrésis, or by the presence of high-value special crops, such as beet in the north, mulberries in the middle Rhône, or the olive around Draguignan.
None the less, more than 60 per cent of the total value of
production was accounted for by arable and special crops
throughout Brittany, the Campagne de Caen, the Lower Seine,
Picardy, Artois, Flanders, Dauphiné and the middle Rhône.
Only in a dozen arrondissements was less than 30 per cent of the
total value generated in this way, largely because of the signifi-

cance of viticulture in Gironde and Aube and of pastoral farming in Cantal, the Pays d'Auge and parts of the Pyrenees.

The values for each of these five categories have been compared according to the modified Weaver technique to produce a combination map (Figure 12.1). The picture is one of surprising uniformity over most of France, where the structure of financial yield was best described by the two-element 'arable and livestock' model. Literal symbols indicate where a three-element model gave the best fit, with vines, firewood and grass entering the description. Only four arrondissements were characterised as 'arable, etc.' (i.e. including special crops) and eight others in Languedoc, Provence, Beaujolais, Charente and Gironde were best described as 'arable and wood and livestock'. The label 'arable and wood and livestock' fitted the single arrondissement of Sartène in southern Corsica. Vines led the financial yield in only six arrondissements, of which three were in Gironde. Finally, livestock contributed the largest proportion in the total value of production in Cantal, northern parts of the Massif Central and its margins in Berry, Bourbonnais and Nivernais, the Vosges and a number of widely dispersed locations including the Pays d'Auge, parts of the Pyrenees, and the urbanised arron-

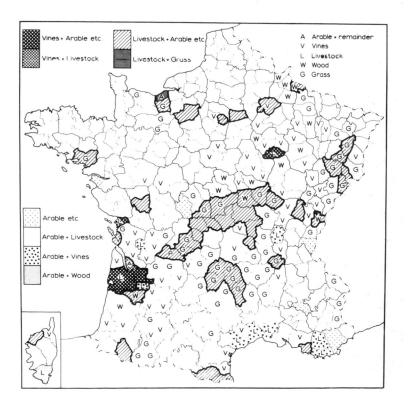

Figure 12.1: Value Combination Map

dissements of Marseilles and Saint-Etienne and the département of Seine.

The overwhelming importance of arable production has been demonstrated once again but, of course, animal rearing did form an essential aspect of all types of arable husbandry and the national table of production values may be adjusted to take that into account. Artificial meadows, fallows, many sections of pâtis and the greater share of oats provided feed for livestock and when the values for these components are added to those for natural grassland and livestock the value of pastoral activities is enhanced to 47.67 per cent of the total (Table 12.1). When allowance is made for the estimated amounts of other grains that were fed to livestock (see Table 3.1) the proportion of total value of production associated with animal husbandry rises to approximately one half.

The value of production in each arrondissement as specified in the Statistique reflects the amount of land under the various crops grown, the productivity of those groups, the numbers of livestock kept, and the prices that each of those commodities commanded. Thus a predominance of generally low-value crops, such as buckwheat or maize, would tend to depress the arrondissement total, while the presence of high-value crops, such as mulberries or market gardening, would enhance it. Areas in which commodity prices were above average would be characterised by higher values of production/ha and vice versa. Fertile soils and intensive farming systems would boost productivity and also serve to raise the value of production. Prior to the construction of an efficient national transport system, areas with high prices (which reflected powerful demand) and/or high productivity (which reflected fertile soils and large inputs of labour) would be found close to major demand centres or in locations that were served by navigable waterways or well-maintained highways. Agriculture would take on its most commercial form in such areas, with producers being spurred on to experiment and innovate by the presence of growing urban demands for food. In such nodes, with high values of production/ha, one would expect new crops to be introduced and new agricultural techniques to be tried, although if labour was cheap and plentiful and properties highly fragmented early attempts at mechanised agriculture might not be expected. New machines would be experimented with in less intensively farmed and settled areas which surrounded these nodes. By contrast, low productivity and/or low commodity prices would characterise areas that were beyond the spread-effects generated by consumers living in large cities. They would be poorly served by transport services and their farming population would tend to cling to traditional, semi-subsistent modes of production.

The value of production quoted in the Statistique exceeded 200 F/ha in some twenty-eight arrondissements, of which nine were in the Ile-de-France, eleven in Nord and Pas-de-Calais, and four in Upper Normandy, with more distant examples in Lower Normandy (Caen and Cherbourg) and around Marseilles and Strasbourg (Figure 12.2). The very highest values surrounded the capital (Paris, 3,375, Saint-Denis 575, Sceaux 463), the northern cities of Lille (394) and Douai (307), and Versailles (305). Values of production were extremely high throughout Nord with values for four of its five remaining arrondissements (Cambrai, Dunkirk, Hazebrouck and Valenciennes) being between 286 and 292 F/ha. The arrondissement of Meaux in Brie (297) was in the same league but values were lower in other parts of the Ile-de-France (e.g. Pontoise 258, Corbeil 249) with a line of territory to the south of Paris extending from Dreux to Coulommiers registering figures below 200 F/ha. The heart of the granary of Beauce around Chartres yielded 210 F/ha and was exceeded in the Pays de Caux (Le Havre 263) and in Roumois (Pont-Audemer 231).

These core areas of commercial farming in the Ile-de-France, Normandy and the north were linked by an arc of arrondissements yielding between 175 and 200 F/ha, which ran from

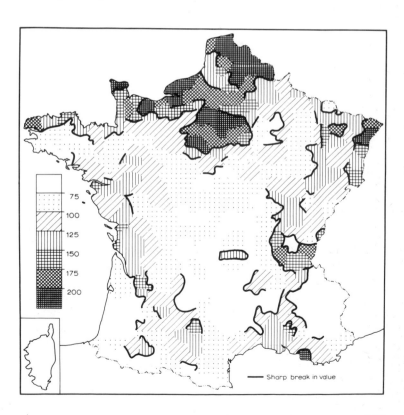

Figure 12.2: Value of Output (F/ha)

Rouen through the Pays de Bray, Beauvaisis and the Oise valley. Yields in excess of 150 F/ha encompassed a broad region which extended virtually without interruption from Cherbourg to Avesnes and from Dunkirk to Chartres. Only three arrondissements in Somme fell below that threshold but lower values around Les Andelys and Evreux separated the Ile-de-France from Normandy. Very few pays elsewhere in France achieved such high values of production. These included the agricultural hinterlands of major provincial cities (with Lyonnais and Dauphiné being the clearest example); intensively cultivated lowlands (such as Alsace, the Moselle valley and the environs of Avignon); *pays viticoles* in Bordelais, around Gaillac and in the pays charentais; and heavily fertilised areas, such as Léon and Trégor, which despatched cereals and vegetables for internal and external trade. The seventy-four arrondissements that yielded over 150 F/ha covered 14.5 per cent of the country, contained 26.0 per cent of its population and provided 25.4 per cent of the national value of agricultural production. In general terms, these were the dimensions of the truly progressive sections of rural France in the 1830s.

Values were also above average in many of the remaining arrondissements in north-eastern France, the northern section of the Paris Basin, Perche, the Lower Loire, Véndee, the pays charentais, the Garonne valley and the Mediterranean coastlands. By contrast, output was valued as less than 75 F/ha in five well-defined regions: the Alps, the Landes, Corsica, most of the Pyrenees, and the greater part of the Massif Central except the Limagnes and Velay. Popularly recognised mauvais pays were reflected by low values in individual arrondissements. These included the Sologne (Romorantin 49), the Morvan (Château-Chinon 63), the Dombes (Trevoux 74), Arcoët, (Loudéac 73), Champagne pouilleuse (Arcis 63) and the Brennes (Le Blanc 72). Twelve arrondissements in Corsica, the Alps and Lozère registered values below 40 F/ha with the absolute minimum around Calvi (17).

High-value commercial farming involved a northern multi-nuclear region, an eastern axis from Alsace to Marseilles, and a less pronounced western axis from the mouth of the Loire to the Garonne valley (Figure 12.2). Sharp breaks in value separated the pays of the Ile-de-France from their less commercialised agricultural surroundings beyond a radius of roughly 80 km from the capital. Converging natural routeways, a dense road network, favourable environmental conditions for arable farming and 1,882,000 consumers in Seine, Seine-et-Oise and Seine-et-Marne made the 1,171,300 ha that they contained the most commercially orientated agricultural region of France. The Oise valley allowed agricultural produce to be brought from areas to the north of the capital and accounted for the gentler gradation of values in that direction compared with the sharp contrast between the Ile-de-France and pays located to the east, south and west. For the farmers of Oise Paris was 'an insatiable giant . . . a monster that paid generously for all that it was given to devour'.[4]

The north (Nord and Pas-de-Calais) was of similar size (1,223,500 ha) and population (1,681,000) to the Ile-de-France but included a cluster of cities rather than just one large demand centre as in the case of Paris. It was in this densely occupied countryside of small farms that cultivation was held in more esteem than in any other part of France. The superiority of Flemish farming that had been praised by Arthur Young was clearly evident half a century later, with 'the beauty of harvests' becoming noticeably greater several leagues south of Arras.[5] It was as if Flanders was 'better cultivated than gardens in other pays'.[6] The value of production and the intensity of yields declined abruptly between the north and both Picardy and Laonnais. Yields were fabulously high by French standards and according to George Houston might be attributed to six factors: an abundant and judicious use of many types of manure; frequent digging of the most intensively farmed areas; complete extirpation of weeds; regular and repeated hoeing; careful choice of seed; and the pursuit of sensible rotations.[7] He explained that 'the less informed attribute this uninterrupted succession of harvests to the inexhaustible fertility of the soil; but intelligent and well-informed travellers attribute it . . . to the indefatigable industry of the inhabitants and to a highly improved mode of cultivation'. Systems of cultivation were so effective that high yields were obtained even from what would appear to be indifferent soils. However La Flandre Agricole warned its readers not to be self-satisfied with their success and to learn from techniques in use in England and Germany so that their prosperity might be increased even more.[8]

Normandy formed a westerly extension of high-value output, serving its own major demand centre, in the form of the city of Rouen (population 92,000) and the densely-peopled industrial towns and villages of the Lower Seine, and also having easy access to the capital. The northern margins of Perche provided a sharp break in financial yields. The north coast of Brittany might be seen as a further element of this northern region of commercial farming but values of production were substantially lower than in other parts of northern France.

The eastern axis extended from the Rhine and the Moselle along the Rhône and Saône to the Mediterranean. Values of 223 F/ha characterised the immediate surroundings of Strasbourg where cultivation had achieved a high degree of perfection but they fell sharply toward the Vosges and declined more gradually in Haut-Rhin. To the west, Nancy (169) and the middle Moselle (Châteausalins 184) formed a second core of high values, which stood in abrupt contrast with Haute-Marne. Values of 125 to 150 F/ha were recorded in the pays of the Saône valley which despatched grain and wine to Lyons, although much of the city's food supply was drawn from Dauphiné to the immediate southeast. These fertile lowlands contrasted with the marshy Dombes, the Alps and the eastern fringes of the Massif Central where farming yielded much less, although Velay provided foodstuffs for Saint-Etienne and Lyons. Avignon and Marseilles formed the remaining foci of high-value production along the eastern axis which was sharply differentiated along much of its length by upland terrain to both east and west.

The western axis extended through a range of coastlands and valleys from Nantes to the middle Garonne valley. Its four core areas, around Paimboeuf, La Rochelle, Libourne and Gaillac, were smaller and generated lower values than those along the eastern axis, with none producing more than 175 F/ha. Unlike the sharp breaks in value around the northern region and along many sections of its eastern counterpart, most of the western axis merged gradually into neighbouring pays although there were quite sharp differences in value between Vendée and Upper Poitou and between the pays vignerons of Bordelais and the Garonne valley and their inferior surroundings in the Landes and the western fringes of the Massif Central.

This alternative view of French agriculture overcomes the difficulties inherent in topical selectivity and spatial generalisation that were encountered in Morineau's analysis.[9] It shows clearly that at this stage in her economic development France was composed of a series of fairly discrete functional hinterlands and axes that were separated by wide areas where agricultural production was not geared primarily to meeting the needs of urban markets. The fiscal promises of the 1790s were far from being realised in the 1830s since France still failed to function as a unified economic entity.

Plus Ça Change . . .

The northern region and the eastern and western axes owed part of their privileged status in French agriculture to the facts of physical geography. The plains, valleys and basins that they comprised were all relatively low lying, enjoyed more equable climates and contained more inherently fertile soils than the broad regions beyond. Their topography made them the 'natural routeways' of France along which settlement proliferated and great cities flourished for purposes of administration and specialist trade, but to a large extent their growth and survival depended on the quantities of food that could be produced in their immediate surroundings. Soils were fertilised heavily, special high-value crops were raised with skill, and the profits derived from supplying urban consumers were returned to the land. The unique political history of France established Paris as the national capital and focus of an administrative system which became increasingly centralised after 1789. Territorial gains during the seventeenth and eighteenth centuries added Flanders and Alsace to the French realm and colonial and maritime policies enhanced the status and activity of the western ports and their agricultural hinterlands which exported not only high-value commodities, such as wines and eau de vie, but also essential supplies of grain. Growing demand and piecemeal improvement in internal communications allowed these farming hinterlands to expand slowly but not until the age of the railways could the centuries-old pattern be gradually changed and more distant pays be brought into the orbit of commercial farming.

Market-orientated areas were exceptions to the rule in France during the July Monarchy, when one-fifth of the land was still moorland or rough grazing, no less than half of the surface was devoted to arable farming and average seed:yield ratios for wheat (1:6.1) and rye (1:5.4) were miserably low. Over a quarter of the arable realm was fallowed each year and only 6.5 per cent of ploughland was devoted to artificial meadows. The poor harvests that followed bad weather continued to cause distress and were to give rise to panic in the 1840s as they had done so many times in the past since supplies just could not be moved with sufficient ease to relieve hardship in areas of food shortage. A dry spring and heavy summer rains in 1845 produced very low yields of all three major bread grains (wheat, rye and maslin) whose collective harvest plummeted from c. 120,000,000 hl/per annum to 113,000,000 hl in 1845 and only 98,000,000 hl in 1846.[10] In addition, the potato crop was affected by disease (*phytophtora*) and production plunged from c. 103,000,000 hl/per annum to 78,000,000 in 1845 which was below the average figure for the earliest years of the July Monarchy.[11] General conditions were even worse in 1846 but buckwheat and maize produced good harvests which were a quarter above average and helped to reduce the measure of distress experienced in regions where they entered into the popular diet. Bread prices increased throughout the country, reaching a peak in parts of western France where the administration had to take action against marauding bands of beggars.[12] The spectre of famine had reappeared and confirmed that the *ancien régime économique* still lived on, more than half a century after 1789. The July Monarchy came to an end in such circumstances of hunger, distress and widespread unemployment because of industrial crisis.

In the following half-century the emerging railway network was to perform a space-shrinking role which gradually allowed distant pays to specialise their agricultural activities. But to a great extent the foci of high-value agricultural production retained their long-established advantages, related to skill, experimentation and capital investment. These were the areas where farming 'had reached the state of science' whilst it remained 'bound by the most detestable routine elsewhere.'[13] It was in these favoured regions that improved ploughs outnumbered traditional *charrues du pays*, early forms of agricultural machinery were purchased, land was first underdrained, and wide expanses of soil were enriched through the application of artificial fertilisers as the nineteenth century progressed.

Material conditions were different elsewhere and other, more basic types of challenge had to be faced. The improvement of rural roads was a prime necessity in many parts of the country but especially in the Massif Central and the west. Growing concern about water management followed widespread flooding in 1846 and the administration established a hydraulic service in each département in 1848 as a branch of the Ponts-et-Chaussées to examine marshes and waterlogged land and to devise drainage schemes.[14] During the Second Empire special funds were made available for improving the 'interior colonies' of the Sologne, Dombes, Bresse, Camargue, Landes and the plains of Forez and eastern Corsica. Action was started for draining marshes in these mauvais pays, bringing some sections into cultivation and putting extensive areas under trees. Ravaged mountainsides were reafforested in southern and central France by the Service des Eaux-et-Forêts and wide stretches of *savarts* on poor calcareous soils in the eastern Paris Basin that proved unresponsive to fertilisation for cropping were also put under trees. Water control was improved in the marshlands of western France and extensive areas of moor and heath were reclaimed for cultivation in Armorica, the Massif Central and middle France as railway lines, special roads and even agricultural canals allowed lime and other fertilisers to be transported to upgrade acid soils.[15] Many of these changes were beyond the immediate realm of agriculture and défrichement was in no way synonymous with the diffusion of commercial farming.

Even more important than differences in space were fundamental differences in outlook between countryfolk living around gradually expanding nodes and axes of commercial agriculture and their compatriots in the backwoods. Centuries of tradition and relative self-sufficiency were not to be forgotten in remote parts of France simply because a railway line had been built and a station opened in the nearest market town. Minds had to be opened to new opportunities such as the possibility of borrowing money from the *Crédit Foncier* which was established as early as 1852 but had very little appeal to the operators of small and medium-sized family farms. Such enlightenment could only be achieved by a programme of general education. A start had been made during the July Monarchy with the *Loi Guizot* of 1833 obliging every commune or group of communes to operate a school and each group of départements to have a college for training teachers.[16] It was heralded as 'the first charter of primary education' and certainly reduced the proportion of conscripts unable to read and write from 50 per cent in 1833 to 39 per cent in 1850, but standards were usually low and attendance perfunctory as children were kept away to work on the family farm during busy periods of the year.[17] Primary education was not to become free, obligatory and universal until legislation in 1881 and 1892.

For rising numbers of Frenchmen the new opportunities that unfolded before them involved migrating to the towns and cities of France but even so, many retained strong attachment to their pays nataux. For others the new opportunities involved directing at least part of their farming activities toward meeting the requirements of urban demand centres and it was this section of the farming population that contributed most to the changes that French agriculture underwent during the half-century after Louis-Philippe's *Statistique*. The arable surface remained remarkably constant over that period, continuing to occupy half the country, but heaths and moorland contracted by one-quarter whilst meadows, woodlands and vines increased modestly (Table 5.2).[18] Artificial meadows doubled in area and fallows declined by half. In spite of these changes and some striking examples of agricultural specialisation and rising productivity, average cereal yields remained depressingly low in the early 1890s with seed:yield ratios for wheat (1:6.6) and rye (1:6.7) being precious little better than they had been in the 1830s. Old ways lived on among a surprisingly large proportion of countryfolk who for a remarkably long period continued to function on the very margins of commercial agriculture and to centre most of their activities on satisfying the basic and internal requirements of family survival.[19] Owning land of their own and sheltered by tariff barriers from the forces of the world market they continued to follow traditional ways and were only gradually touched by the forces producing modernisation. Until very recently this characteristic encapsulated the uniqueness of the French among their neighbours since farming as a way of life has only been eroded substantially in France during the second half of the twentieth century.

Notes

1. Mme Romieu, *Des Paysans et de l'agriculture en France au XIXe siècle* (Paris, 1865).

2. A. Blanqui, 'Les populations rurales de la France en 1850', *APAPER*, vol. 24 (1851), p. 197; F. Lullin de Châteauvieux, *Voyages agronomiques en France* (Paris, 1843), pp. 85-97.

3. M. Morineau, *Les Faux-semblants d'un démarrage économique en France au XVIIIe siècle* (Paris, 1970).

4. H. Daudin, 'Exposé de l'état de l'industrie agricole dans le département de l'Oise', *Bulletin de la Société Agricole et Industrielle de l'Oise*, vol. 1 (1834), p. 85.

5. C. de Gourcy, *Notes sur l'agriculture des départements du Nord et du Pas-de-Calais* (Paris, 1847), p. 10.

6. Anon., 'Note préliminaire', *La Flandre Agricole et Manufacturière*, vol. 1 (1834-5), p. ii.

7. G. Houston, 'Flemish husbandry', *Letters and Extracts on Agriculture: Charleston, South Carolina* (1825), p. 189.

8. Anon., 'Rapport sur les réponses données par les membres des comices agricoles', *Journal de la Société des Sciences, Agriculture et Arts du Département du Bas-Rhin*, vol. 4 (1827), p. 310.

9. Morineau, *Les Faux-semblants*.

10. E. Juillard (ed.), *Histoire de la France rurale: apogée et crise de la civilisation paysanne, 1789-1914* (Paris, 1976), vol. 3, p. 140.

11. E. Labrousse (ed.), *Aspects de la crise et de la dépression de l'économie française au milieu du XIXe siècle* (Paris, 1956), p. v.

12. R. Laurent, 'Le secteur agricole', in F. Braudel and E. Labrousse (eds.), *Histoire économique et sociale de la France* (Paris, 1976), vol. 3 (2), p. 756.

13. M. Delbert, 'De la petite culture en Bretagne', *JAP*, vol.4 (1852), p. 89.

14. L. Girard, *La Politique des travaux publics du Second Empire* (Paris, 1952), p. 31.

15. K. Sutton, 'The reclamation of wasteland in the Sologne', *TIBG*, vol. 52 (1971), pp. 129-44; K. Sutton, 'A French agricultural canal: the Canal de la Sauldre', *AHR*, vol. 21 (1973), p. 51-6.

16. R.G. Anderson, *Education in France, 1840-70* (Oxford, 1975), p. 7.

17. A. Cobban, *A History of Modern France* (3 vols., Harmondsworth, 1965), vol. 2, p. 125.

18. Ministère de l'Agriculture, *Statistique agricole de la France: 1892* (Paris, 1897).

19. G.W. Grantham, 'The diffusion of the new husbandry in northern France, 1815-1840', *JEH*, vol. 38 (1978), pp. 311-37.

Printed versions of a standard questionnaire were despatched to every commune in France. The returns were then checked and summated for each arrondissement. The published version of the Statistique was printed unevenly on rough textured quarto sheets which were bound together tightly and which rendered extensive photocopying impractical. Numerical data were therefore extracted systematically for the present study and were transferred manually on to coding forms prior to being punched on to 80-column data cards for computerised tabulation, classification and display. Each arrondissement and each batch of information (for example, a single crop or a group of related crops or types of livestock) was given a 3-digit numerical label to assist identification and to facilitate mechanical retrieval. These numerical descriptions were retained throughout. Many quantitative maps were plotted manually and drawn in conventional ways. Indeed, traditional techniques were the most appropriate for 'combination' maps which made use of wide ranges of shading and large numbers of unevenly located literal symbols. By contrast, automated techniques that make use of data files stored on disk were employed successfully for feeding relatively restricted ranges of line shadings and proportional symbols on to a consistent spatial base that had been digitised from a trace of administrative areas. Unlike many computergraphic processes, the map output was not generated as ink on paper but was etched on to microfilm from which photographic prints were made in the normal way.

This particular form of automated cartography offers at least five major advantages.

1. It generates a high-quality end product with a very acceptable standard of legibility, unlike many earlier computergraphic techniques.

2. It allows a very large quantity of numerical data to be displayed rapidly, in a fraction of the time that it would take a cartographer to produce comparable line drawings.

3. Mechanical calculation, construction and plotting of proportional symbols and various densities of linework spare the cartographer the burden of these extremely laborious tasks. Using very large numbers of proportional symbols would not really prove feasible if human rather than mechanical labour was employed.

STATISTIQUE DE FRANCE.

POPULATION DE LA COMMUNE:

habitants.

SON ÉTENDUE :

hectares.

DÉPARTEMENT d

TABLEAU des Cultures de la Commune, de leur étendue et de leurs produits.

NATURE des CULTURES ET PATURAGES.	ÉTENDUE de CHAQUE CULTURE en hectares.	PRODUIT TOTAL			PRIX MOYENS EN FRANCS ET CENTIMES			QUANTITÉ DE SEMENCES, en kilogrammes ou en hectolitres.	QUANTITÉ de la CONSOMMATION totale en kilogrammes ou en hectolitres.
		en KILOGRAMMES.	en HECTOLITRES.	en STÈRES.	du KILOGRAMME.	de L'HECTOLITRE.	du STÈRE.		
					fr. cent.	fr. cent.	fr. cent.		
Froment............................				,			,		
Méteil............................				,			,		
Seigle............................				,			,		
Orge............................				,			,		
Avoine............................				,			,		
Sarrasin............................				,			,		
Maïs ou millet....................				,			,		
Pommes de terre.................				,			,		
Légumes secs....................				,			,		
Betteraves........................				,			,		
Vignes... { Vins...............				,			,	,	
Vignes... { Eau-de-vie...........				,			,	,	
Prairies artificielles.............				,			,		
——— naturelles.............				,			,	,	
Bois............................								,	
Jachères........................				,			,	,	
Jardins........................				,			,		
Colza, navette (huile)............				,			,		
Oliviers (huile)..................				,			,	,	
Tabac............................				,			,		
Pastel............................				,			,		
Lin............................				,			,		
Chanvre........................				,			,		
Mûrier (soie)....................				,			,	,	

Version of the questionnaire despatched to each commune in France.

QUANTITÉS ET VALEURS DU BÉTAIL, DES TROUPEAUX, CHEVAUX ET AUTRES ANIMAUX UTILES
EXISTANT DANS LA COMMUNE.

	TAUREAUX.	BŒUFS.	VACHES.	VEAUX.	TOTAL du BÉTAIL.	BÉLIERS.	MOUTONS.	BREBIS.	AGNEAUX.	TOTAL des TROUPEAUX.	PORCS.	CHÈVRES.	CHEVAUX.	JUMENTS.	POULAINS.	MULES et MULETS.	ÂNES.
	Nombre.	Nombre.	Nombre.	Nombre.	Nombre.	Nombre.	Nombre.	Nombre.	Nombre.	Nombre.	Nombre.	Nombre.	Nombre.	Nombre.	Nombre.	Nombre.	Nombre.
Quantité de chaque sorte d'animaux....																	
Valeur moyenne de chacun d'eux....	fr. c.	fr. c.	fr. c.	fr. c.	fr. c.	fr. c.	fr. c.	fr. c.	fr. c.	fr. c.	fr. c.	fr. c.	fr. c.	fr. c.	fr. c.	fr. c.	fr. c.
Revenu moyen donné par chacun d'eux....																	

CONSOMMATION DE LA VIANDE.

	BŒUFS.	VACHES.	VEAUX.	TOTAL du BÉTAIL.	BÉLIERS.	MOUTONS.	BREBIS.	AGNEAUX.	TOTAL des TROUPEAUX.	PORCS.	CHÈVRES.
Nombre d'animaux de chaque sorte abattus annuellement....	Nombre.	Nombre.	Nombre.	Nombre.	Nombre.	Nombre.	Nombre.	Nombre.	Nombre.	Nombre.	Nombre.
Poids { brut.... de chacun d'eux en net.... } kilogrammes.	kilogrammes.	kilogrammes.	kilogrammes.	kilogrammes.	kilogrammes.	kilogrammes.	kilogrammes.	kilogrammes.	kilogrammes.	kilogrammes.	kilogrammes.
Quantité totale de la viande consommée....											
Prix moyen du kilogramme de chaque espèce de viande....	fr. c.	fr. c.	fr. c.	fr. c.	fr. c.	fr. c.	fr. c.	fr. c.	fr. c.	fr. c.	fr. c.

OBSERVATIONS.

4. New variables can be created within the machine (for example, determining the difference between production and consumption of a selected commodity) and then plotted automatically.

5. Many trial maps can be easily prepared to help the researcher assimilate large quantities of spatially distributed statistics. It is not necessary for prints to be made in every case since trial maps can be inspected on a microfilm reader and, indeed, only a very small proportion of the computergraphics produced as part of the present exercise have been included in this volume. A wider range is to be found in H.D. Clout, *Agriculture in France on the Eve of the Railway Age* (PhD thesis, University of London, 1979) which also contains a much fuller record of publications relating to local aspects of agricultural activity in France during the first half of the nineteenth century.

Angeville, D' A. *Essai sur la statistique de la population française* (Paris, 1836, repr. 1969)

Armengaud, A. *Les Populations de l'Est aquitain* (Paris, 1961)

Aron, J.P., Dumont, P. and Le Roy Ladurie, E. *Anthropologie du conscrit français* (Paris, 1972)

Auge-Laribé, M. *La Révolution agricole* (Paris, 1955)

Baudrillart, H. *Les Populations agricoles de la France*, 3 vols. (Paris, 1885, 1889, 1893)

Béteille, R. *Les Aveyronnais* (Poitiers, 1974)

Blanqui, A. 'Les populations rurales de la France en 1850', *APAPER*, vol. 24 (1851), pp. 189-220

Block, M. *Statistique de la France*, 2 vols (Paris, 1860)

Bozon, P. *La Vie rurale en Vivarais* (Clermont-Ferrand, 1961)

Brunet, P. *Structure agraire et économie rurale des plateaux tertiaires entre la Seine et l'Oise* (Caen, 1960)

Brunet, R. *Les Campagnes toulousaines* (Toulouse, 1965)

Caron, F. *An Economic History of Modern France* (London, 1979)

Cavaillès, H. *La Route française* (Paris, 1946)

Chombart de Lauwe, J. *Bretagne et Pays de la Garonne* (Paris, 1946)

Contamine, H. *Metz et la Moselle de 1814 à 1870*, 2 vols. (Nancy, 1932)

Duby, G. and Wallon, A. (eds.), *Histoire de la France rurale*, vol. 3 (Paris, 1976)

Dutil, L. *L'Etat économique du Languedoc à la fin de l'ancien régime* (Paris, 1911)

Faucher, D. *Géographie agraire: types de cultures* (Paris, 1949)

Fel, A. *Les Hautes Terres du Massif Central français* (Paris, 1962)

Festy, O. *L'Agriculture pendant la révolution française* (Paris, 1947)

Frémont, A. *L'Elevage en Normandie*, 2 vols. (Caen, 1967)

Garrier, G. *Paysans du Beaujolais et du Lyonnais, 1800-1970*, 2 vols., (Grenoble, 1973)

Gille, B. *Les Sources statistiques de l'histoire de la France des enquêtes du XVIIIe siècle à 1870* (Paris, 1964)

Hemardinquier, J.J. *Pour une histoire de l'alimentation* (Paris, 1970)

Hohenberg, P.M. 'Maize in French agriculture', *JEEH*, vol. 6 (1977), pp. 63-101

Houssel, J.P. (ed.), *Histoire des paysans français* (Roanne, 1976)

Husson, A. *Les Consommations de Paris* (Paris, 1856)

Jardin, A. and Tudesq, A.J. *La France des notables, 1815-48*, 2 vols. (Paris, 1972)

Jorré, G. *Le Terrefort toulousain et lauragais* (Toulouse, 1971)

Labrousse, E. (ed.), *Aspects de la crise et de la depression de l'économie française au milieu du XIXe siècle, 1846-51* (Paris, 1956)

Laurent, R. *Les Vignerons de la Côte-d'Or au XIXe siècle*, 2 vols. (Dijon, 1958)

——, *L'Octroi de Dijon au XIXe siècle* (Paris, 1960)

——, 'Le secteur agricole' in F. Braudel and E. Labrousse (eds.), *Histoire économique et sociale de la France* (Paris, 1976)

Léonce de Lavergne, G. *Economie rurale de la France depuis 1789* (Paris, 1861)

Le Play, F. *Les Ouvriers européens*, 6 vols., (Paris, 1879)

Lerat, S. *Les Pays de l'Adour* (Bordeaux, 1963)

Leuillot, P. *L'Alsace au début du XIXe siècle*, 3 vols. (Paris, 1959)

Lullin de Châteauvieux, F. *Voyages agronomiques en France* (Paris, 1843)

Morineau, M. 'Y-a-t-il eu une révolution agricole en France au XVIIIe siècle?' *RH*, vol. 486 (1968), pp. 299-326

——, 'Prix et "révolution agricole"', *AESC*, vol. 24 (1969), pp. 401-23

——, *Les Faux-semblants d'un démarrage économique en France au XVIIIe siècle* (Paris, 1970)

——, 'La pomme de terre au XVIIIe siècle', *AESC*, vol. 25 (1970), pp. 1767-85

——, 'Révolution agricole, révolution alimentaire, révolution démographique', *ADH* (1974), pp. 335-71

Newell, W.H. 'The agricultural revolution in nineteenth century France', *JEH*, vol. 33 (1973), pp. 697-731

O'Brien, P.O. and Keyder, C. *Economic Growth in Britain and France, 1780-1914* (London, 1978)

Price, R. *The Economic Modernisation of France, 1730-1880* (London, 1975)

Renucci, J. *Corse traditionnelle et Corse nouvelle* (Lyons, 1974)

Sée, H. *La Vie économique de la France sous la monarchie censitaire, 1815-48* (Paris, 1927)

Sigaut, F. 'Pour une cartographie des assolements en France au début du XIXe siècle', *AESC*, vol. 31 (1976), pp. 631-43

Thuillier, A. *Economie et société nivernaises au début du XIXe siècle* (Paris, 1974)

Toutain, J.C. 'Le produit de l'agriculture française de 1700 à 1958', *Cahiers de l'ISEA*, vol. 115 (1961), Série AF1, pp. 1-221

Vidalenc, J. *Le Département de l'Eure sous la monarchie constitutionnelle, 1814-48* (Paris, 1952)

——, *Le Peuple des campagnes: la société française de 1815 à 1848* (Paris, 1970)

Weber, E. *Peasants into Frenchmen: the modernization of rural France, 1870-1914* (London, 1977)

Young, A. *Travels during the years 1787, 1788 and 1789* (London, 1792)

Zeldin, T. *France, 1848-1945*, 2 vols, (Oxford, 1973, 1977)

In addition, virtually all of the geographical monographs published before 1960 contain useful regional information.